Design For Reliability

Developing Assets That Meet The Needs of Owners

Daniel T. Daley

Industrial Press Inc.

Library of Congress Cataloging-in-Publication Data

Daley, Daniel T.
  Design for reliability / Daniel T. Daley.
    p. cm.
  Includes bibliographical references and index.
  ISBN 978-0-8311-3437-2 (hard cover)
  1. Reliability (Engineering) 2. Manufactures--Quality control. I. Title.

TS173.D35 2011
658.2'7--dc22

2010048943

Industrial Press, Inc.
32 Haviland Street, Suite 3, South Norwalk, CT 06854
Tel: 203-956-5593, Toll-Free: 888-528-7852
Email: info@industrialpress.com

Sponsoring Editor: John Carleo
Copyeditor: Robert Weinstein
Interior Text and Cover Design: Janet Romano

books.industrialpress.com
ebooks.industrialpress.com

10   9   8   7   6   5   4   3   2   1

Dedication

I have an eight year old grand-daughter who is deaf and blind.
If you were to watch her for any period of time, you would notice
that she is continually moving around trying to do things.
I also have a ninety-one year old mother who, in common with my
grand-daughter, is always busy trying to do things.
I am inspired by them both.
This book is dedicated to all the people who ignore limitations
and just keep trying.

Table of Contents

Introduction

Some time ago, after years of dissatisfaction with the reliability of new assets that were being purchased, the company where I worked decided to demand Design For Reliability (DFR) as a standard part of the design process. We had some of our own ideas of what we wanted from the companies performing the designs for new assets. However, we decided to start by asking the suppliers for a description of their current design process used to ensure adequate reliability of their finished product.

In response, they provided us with a litany of different forms of reliability analysis. After spending some time studying their responses, one of our reliability engineers made the following observation: "It looks as though they Googled the word reliability and wrote down everything they found." His point was that there are myriad analytical techniques for reliability; simply applying a significant number of them does not result in an effective DFR process.

Based on the reliability performance of the systems we had been receiving in past years, we suspected that the processes our suppliers were using were not really comprehensive and integrated DFR processes—at least not from the perspective of the future owner. After a lot of head scratching, we realized that we were probably facing at least two problems:

1. Were suppliers using any effective form of DFR?
2. If suppliers were using some form of DFR, was the process being used intended to provide a product that met our (the owner's) needs?

Because we were dealing with several entities that provided us with systems, we knew that the different suppliers developed their products in different ways. We were convinced that the method used by one of the suppliers was better than the others. But this did not lead to better performance for us (the long-term owner)—only a different kind of performance. The business approach or business model was different for each supplier. One provided a product and then seemed to turn their back on it as soon as the warranty period was over. Another supplier was quite the opposite. When you purchased a product from that supplier, it seemed as though you were "marrying" the supplier because you spent the entire life of the asset bound to the supplier as a source of parts or service.

These differences in the suppliers and in asset performance that seemed to match the supplier's business model made us realize that products were being designed to the seller's business model. Along with that realization came the epiphany that suppliers were designing and building products to their business model, not ours.

This point is important to understand. Most sellers say in their advertising that they are building their products for their customers. They even go so far as to create devices like the "Voice of the Customer," a survey technique for gathering feedback that makes customers believe their needs are being addressed. Certainly sellers are interested in addressing their customer's concerns—as long as those concerns are consistent with the seller's business model. Quite often customers have suggestions that are good for both the seller and the future owner. In these cases, the seller is willing to address the owner's needs.

I understand that my point of view may sound cynical. Even the most naïve individual must realize that what we are discussing is "only" business. A simple definition of the term business model is as follows: A business model describes the rationale of how an organization creates, delivers, and captures value. All successful companies have a well-designed business model. They survive and prosper by following their business model. If the business model contains shortcomings that limit or prevent their success, they will change the business model to enhance their success. Many elements of the business model have to do with their products.

Once products are sold and out the door, reliability tends to be a drag on future business. If their products are too reliable, the owner will have no reason to maintain or replace them. For the asset to be a source of future business, there either needs to be new features to attract the owner back or the asset needs to begin performing more poorly as it ages. There is no assurance that new features will drive repeat business so some part of the future business must be driven by the desire to renew the reliability that was provided when the asset was new. At least this is the case that is driven from the perspective of the seller's business model.

As an owner of an asset, the point of view is entirely different. An owner who purchases an asset wants it to last forever and to be reliable for as long as it lasts. The only reason an owner may want to purchase another asset is that the business has expanded and more production capacity is needed.

The business objectives identified in the business models ultimately find their way into the design process for assets being either sold or purchased. The seller's business model will dictate the elements of the Design For Reliability (DFR) process used by the seller. If long-term owners want the elements of their business model to be at the center of the DFR process used to develop the assets they purchase, the owners will need to provide purchase specifications that ensure their needs are met. They will also need to be willing to pay additional costs to provide resources to perform the analysis that ensures their needs are met during design and construction. Additional resources are also needed to collect information and publish the reports showing if their objectives are being met after delivery.

In the past, most purchase decisions were made exclusively based on first cost. The least expensive asset was typically selected. In the relative recent past, companies have begun to focus on Life Cycle Costs (LCC) or the Total Cost of Ownership (TCO). These systems of describing costs take into account the first costs; they also include the costs for maintaining the asset over its entire life and the cost of lost production when the asset is not available to perform its intended function. Although it would be naïve to think that all companies and all individuals within companies have accepted the logic of using LCC or TCO analysis to help long-term thinking, more and more people are taking the long view of asset costs and value.

The only way to make a case for a decision based on LCC

or TCO considerations is to perform an analysis that identifies the costs associated with the various alternatives that are available. The only way to know the future costs of failures and the maintenance needed to either prevent those failures or recover from them once they have occurred is to perform reliability analysis during the design process. The reliability analysis begins by identifying requirements for the asset; it then determines which of the available alternatives will meet requirements at the least overall cost over the entire life of the asset. This reliability analysis that is conducted during the conventional asset design process is called Design For Reliability or DFR.

There are currently a number of books that describe the DFR process from the perspective of the seller (or the entity that designs, builds, and distributes the asset). After many years of attempting to use the products of that analysis to meet the needs from the perspective of an owner, I have decided to write this book that highlights the difference, identifies the owner's needs, and describes how owners should go about seeing that their objectives are addressed during the design and development of the asset.

Clearly, there are some differences in objectives between the seller and the long-term owner. Forcing sellers to perform analysis that supports the owner's needs may not be easy. Some sellers do not use a comprehensive DFR process. Some sellers who do use a DFR process are proud of their system and are unwilling to change. The most sophisticated sellers will recognize the fact that making the changes needed to achieve the owner's objectives may, at the same time, go a long way in preventing them from achieving their own objectives. In any of these cases, it may be a struggle to get the seller to change.

As a result, it is critical that owners be sensitive of the "golden rule." It is they who have the gold who make the rules. In this case, the golden rule applies only before the purchase agreement is signed. It will be possible to get sellers to agree to incorporate the owners' DFR requirements in the product design process only before the purchase agreement is signed. After the owners have agreed to accept the sellers' current product, it is an uphill battle to make meaningful changes.

There are a wide variety of ways in which the customer and long-term owner can procure an asset from a seller (or entity responsible for designing and building the asset).

♦ Owners and operators of refineries, process plants, and power stations work with outside entities in a variety of configurations to obtain new plant facilities. In these situations, the challenge is convincing those entities to change their normal way of conducting business to deliver the customer's needs. Most often there are a fairly limited number of organizations that provide the kind of assets being sought; changing their approach to design is paramount to "teaching old dogs new tricks."

♦ Owners and operators of heavy mobile equipment for construction, transportation, or other needs typically purchase an asset from which there is a choice of several current models. In this case, the challenge is one of changing one of the current models to meet the customer-owner's long-term needs. In doing so, sellers will no longer be able to pull the design for a current product down from the shelf and earn more income on a past resource investment in a design.

Sellers will need to go "back to the drawing boards" to make a new sale. Sellers risk learning that past products contain flaws for which they may still be responsible. Performing an adequate reliability analysis on an existing product (for which a number of copies are already in operation) may tell the sellers something they would prefer not knowing.

◆ In either of these situations and a myriad of others, there are numerous examples where a component or major element will deteriorate over time and require replacement. The design/development process for those individual components is not as comprehensive as a "start from scratch" case, or one in which a major modification to a current model is required. Still, there are many applicable elements of the DFR process that can and should be applied to ensure long-term reliability is maintained or enhanced as a result of the replacement of those components.

Although the basic philosophies used as a basis for DFR may vary from situation to situation, the objectives for applying the process remain much the same. As an owner, your objectives are to obtain an asset that meets your needs, not those of the seller. To do so, you will need to clearly understand your needs and see that the seller is taking all the steps needed to achieve those needs.

Owners must be assured that sellers have performed the analysis and created the data showing that the product will meet the owner's needs. The products of that analysis do not exist in the form of glossy brochures that the seller distributes to anyone who has an interest in their product. Those brochures contain the details that primarily support the seller's business model. The information owners need comes in the form of reports which result from detailed analysis that specifically addresses the owner's requirements.

The task of stimulating the seller to perform that analysis should not be underestimated. The results of the seller's analysis are likely to tell the owner that it will be necessary to pay more in terms of initial cost to assure the long-term performance required to provide lower LCC or TCO. Because not all members of the owners own organization take the long view, battles can be internal as well as external.

Another important point should be kept in mind. A number of years ago I began my efforts to address reliability, availability, and maintainability as a part of the design process. At that time, a number of individuals made comments to the effect: "We will never be able to afford the improvements in robustness or redundancy needed to significantly improve reliability of the assets being purchased." That is a false belief for several reasons:

1. The changes needed to improve reliability are not as expensive as one might think. They primarily depend on additional engineering, not hardware.
2. Viewing the changes from the perspective of TCO or LCC tends to put the changes in first cost into perspective.
3. If you don't perform the analysis, you will never know.

Just Don't Assume That Reliability is Too Expensive.

Over the past few years, it seems that the quest for "green" products has grown dramatically. Readers should rest assured that products that meet all of their objectives and enjoy long and reliable lives are by far the greenest of all green products. Avoiding the need to expend the energy and resources needed to replace an unreliable or short-lived asset is the most environmentally sound approach.

As the author, my fondest hopes are for two things:

First, I hope you learn something from the book that can be directly applied and produce a tangible benefit. Second, I hope each reader enjoys this book. I have made a concerted effort in this book as well as my previous books to make them easy to read and understandable for the broadest audience possible.

Chapter 1

Differences in Perspective

We are all inclined to judge ourselves by our ideals; others, by their acts.

Harold Nicolson

For a moment, try to put yourself in the position of a business owner. You own a number of capital intensive assets and you use them to manufacture a specific slate of products or to provide a specific slate of services. You are in business for the long haul and, in many ways, your profitability depends on the reliability of your assets.

As part of this mental exercise, think about the elements of your business model. How should those elements be managed to deliver the greatest return for your investors? For purposes of this discussion, we can focus almost exclusively on the manner in which your business

model relates to the capital intensive assets that are used in manufacturing your products or providing your services.

Our Capital Asset-Based Business Model

We can probably simplify this discussion by summarizing the features of our capital asset-based business model into a few bullet points:

- The initial cost of the capital asset we purchase must be fair in comparison to the possible return on investment. This point must apply to the Total Cost of Ownership over the entire 30-year life of the asset.
- The impact of reliability as evidenced in terms of the lost profit resulting from unplanned downtime and the cost of repairs resulting from unexpected outages must be consistent with the initial expectations. (If they are different, we will be misled and we will mislead our investors.)
- The impact of availability–as evidenced by the lost profit resulting from planned downtime and the cost of maintenance during those times?must be consistent with the initial expectations.
- The on-going cost of maintenance needed to retain the expected output and other key elements of performance must be consistent with initial expectations.
- The on-going operating costs for operators, fuel, operating materials, etc., needs to be consistent with the initial expectations.
- The on-going effectiveness and efficiency of the asset's operation needs to be consistent with the initial expectations. In other words, we need to be able to produce at the expected

Throughout this book, we will continually refer to three characteristics, how to quantify them, and how they impact the seller and long-term owner's ability to be successful. These characteristics are: Reliability, Availability, and Maintainability. For the sake of clarity, we will provide a brief definition here.

Reliability—Reliability is a measure of the instantaneous likelihood that a system or device will fail in a specific period of time. In some cases, the period that a device is needed is a called a mission. An important aspect of reliability is the fact that the likelihood of an event when viewed over a large population or a long period of time is much the same as the event itself. For instance, a 10% likelihood of an event costing $100,000 is a risk with a value of $10,000. When viewing the same reliability over a ten-year time horizon, the anticipated cost of the reliability debit is $100,000. The risk of failure due to poor reliability is real money.

Availability—Availability is the ratio of uptime or time the system or device is capable of performing its intended function to the total time the device has been provided for that intended service. The availability debits or unavailability are composed of both planned and unplanned outage periods. Planned outages are those periods of time the asset must be shut down in order to perform the maintenance needed to allow it to perform its intended function in a reliable manner. Unplanned outages are those resulting from poor reliability. The poorer the reliability, the more unplanned outages will occur.

Maintainability—Maintainability is a measure of the ability to restore the inherent reliability of a system or device in a ratable period of time. The inherent reliability of a device is the maximum reliability that can be achieved if the device is operated and maintained in an optimum manner. The inherent reliability of a device is ultimately dependent upon the configuration and the robustness of the components from which it was constructed. A ratable period is a definable and repeatable period. For a device to be truly maintainable, it must be possible both to restore the inherent reliability and to do so in a known and repeatable period of time. Understanding the maintainability of a device before it is constructed requires some form of analysis that identifies the different forms of maintenance that will be necessary over the life of the asset. Once the various forms of maintenance are known, they can be reviewed for maintainability.

rates and do so using only the anticipated amount of raw materials.

♦ The usable life of the asset needs to be consistent with the initial expectations.

♦ All these characteristics need to apply over the entire life of the capital asset.

In other words, all aspects that can affect the profitability of the asset for its entire life need to be specified by the owners and confirmed (or adjusted) by the suppliers at the start of the asset's life. Then the asset needs to operate in the agreed-upon manner. Clearly, there is an inherent assumption that the owners know what is needed to support their business model and specifies those characteristics. If they do not know or have not clearly specified their requirements, inadequate performance during the life of the asset is the owners' fault.

The Buyer's Model

As an owner, for my business to function in a manner consistent with my business model, I need to be able to rely of my capital assets. They need to provide the assumed profit opportunity and they need to cost no more than assumed. When they produce less and cost more, my business will go upside-down and my business plans will be meaningless. Ultimately, my performance will be different than I advertised and I will lose the confidence of the investment community.

The Seller's Model

Now, for a moment, try to put yourself in the position of the seller of the capital assets being provided to the business owner. You are in the business of designing, building, selling, and servicing these capital assets. As your business relates to the capital assets or products in question, your business model is significantly different from the owner's business model. The seller's business model depends on the owner continuing to need to purchase, replace, or maintain assets at a rate that supports the seller's success, but may adversely affect the owner's success.

This point gets to the purpose of this entire chapter. *The seller builds a product to fulfill a different business model than the one that is important to the business owner who purchases the asset. An offshoot of that difference is that the DFR process used to serve the seller's needs is different than the DFR process required to meet the owner's needs. Unless owners are aware of those differences and specify the analysis and resulting information needed to meet their needs, it is unlikely they will receive it. And it is unlikely that the asset will be designed and constructed in a manner that meets the owner's needs.*

Let's continue this discussion by further analyzing the seller's business model.

For purposes of this discussion, let's assume that the seller's role in providing service is a limited one. Let's assume the seller provides servicing for failures resulting from poor design or component failures during the warranty period. Further, at the owner's discretion, the seller can provide servicing on a compensated basis for as long as desired by the owner for the entire life of the asset. In addition, the seller may provide replacement components for the life of the asset as long as the buyer does not find a replacement part that works as well as the OEM part, but at a lower cost. (Generally speaking, the OEM provides replacement components, but does so at a premium cost as compared to the actual manufacturer of those components.)

Once again, let's describing the key features that are part of the seller's business model. This viewpoint is based strictly on the capital asset being manufactured by the seller and being purchased by the owner. Although the term typically used is TCO or Total Cost of Ownership, a more appropriate term in this case is TCS or Total Cost of Selling.

- ◆ The initial cost of the product must be priced to make at least an acceptable return on investment for all resources and effort used in all aspects of making the product. This step is only the starting point for setting the price. Although the asset being discussed is primarily seen as a cost for the owner, it is viewed as an opportunity for the seller. **Sellers can price their products at any level the market will bear.** They can add a premium for their reputation. They can even add a premium for the inefficiencies of their competition. In other words, the ability to produce an asset for less than the competition does not mean manufacturers need to price the asset based on their costs. **They can price their products for the market place.**
- ◆ The seller's business model needs to consider the cost of warranty and the duration of the warranty. Some products are made to survive the warranty period and little more. It is in the seller's interest to provide a product that requires very little warranty return. The immediate costs and the poor public image caused by returns during the warranty period result in sellers typically doing a good job of seeing that their products will be reliable during the warranty period.

◆ The seller's business model must take into consideration the value of risk during any direct warranty period or any extended warranty period. You will seldom see a seller agree to assume responsibility for an extended warranty despite the poor performance during the initial warranty period. The value of this risk is not something built into the profit-model for a given product; sellers are reticent to go back and change the profit picture on past products. They will make corrections and accept the immediate cost, but not allow additional unexpected costs to continue to cloud their future beyond the initial warranty period for a product.

◆ The seller's business model needs to consider the cost of future parts sales. Companies carefully guard the names of manufacturers and models for the components used in the systems they manufacture when it is possible they can have an on-going income stream from selling parts. The amount of effort and, at times, poor customer relationships caused by this secrecy attest to the importance sellers place on this income stream. As a result, it can be safely assumed that components are selected in a manner that feeds the income stream. Parts that last forever provide little downstream value to the OEMs or original sellers. (The difficulty caused by this issue will be made clear later when we discuss the data that owners will demand from sellers. In order for owners to identify components that are not meeting expectations, they will need to be provided with far more information than sellers are frequently willing to divulge about the components that make up their products. As a result, this dispute can become a sticking point.)

◆ The seller's business model needs to consider the opportunity available from future servicing. If the seller?s business is structured in a manner that it provides on-going service, the model will be structured so that products will be designed to enhance the opportunity. As with information concerning parts, information concerning appropriate maintenance (e.g., timing and structure of predictive and preventive maintenance) is something that will be guarded. OEM manuals will tell the owner the work needed to prevent failures and protect the asset during the warranty period. They will not, however, help to prevent long-term parts deterioration so the long term market for parts and service is eliminated.

♦ The seller's business model needs to consider the value of long-term risks associated with injuries or fatalities that might result from a product failure. This concern involves both the cost of legal action and the impact on the seller's reputation. This risk can go beyond the warranty period, but there is a practical limit based on the perceived practical usefulness of the asset. (For instance, it the steering on an antique car fails and someone is killed, it is unlikely that the original maker will either be held liable in court or by public opinion.)

At first, the discussion above may seem to paint an altogether negative picture of the seller's perspective. It is not intended to do so. It is simply intended to point out that sellers and owners are in different businesses. As such, the key elements of their business models are different. Frequently, sellers try to tell owners that their only interest is to take care of their customers. That is perhaps the most misleading part of any sales pitch. It is important to keep the old saying in mind: Buyer beware.

Viewing the difference between the seller's vantage point and the owner's vantage point as they relate to reliability, or more specifically to the way Design For Reliability is approached, there are broad differences. These differences can be summarized as follows.

Seller are interested in:
- ♦ Initial profit
- ♦ Reliability during the warranty period
- ♦ A long-term income stream on parts
- ♦ A long-term income stream on service
- ♦ Transferring total accountability for costs (for failures and losses) to owners as soon as possible

Owners are interested in:
- ♦ Initial cost
- ♦ Ability to achieve long-term reliability
- ♦ Ability to achieve long-term availability
- ♦ Ability to achieve design production and performance for the long-term
- ♦ Ability to achieve quick and effective corrective action when performance is not as advertised

Many companies are both owners and sellers. Their interests as owners are different than their interests as sellers. Some companies want to administer expectations by different rules. They want to view their own assets with a long-term view, but view the assets they sell to others in a short-term manner. That doesn't work.

In the long run, the philosophy concerning the usable life of all assets tends to permeate the attitudes of a company's personnel. If the company views physical assets they sell in a "throwaway" manner, employees will probably view owned assets in the same way. The only rationale I can offer for this assertion is that most individuals are fair minded and personally operate under the rule they learned as a child: "Do unto others as you would have them do unto you."

Chapter 2

DFR for the Seller's Business Model

Mistakes are the portals of discovery.

James Joyce

The first chapter demonstrates that the objectives of the sellers, as they pertain to the asset being sold, are different from those of the buyers or long-term owners. If we accept that the objectives are different, we will also recognize that the asset will be designed differently to meet the owners' objectives than it would be to meet the sellers' objectives. That being said, let's consider how the process of DFR differs to achieve those two ends.

A DFR Process

We will begin this discussion by describing a simplified pattern for completing the design of a new or modified product. It is important to understand the design process because DRF is an activity that must be accomplished concurrently with the product design. The discussion continues by describing a DFR process that is intended to support the seller's objectives. It is important to emphasize the term a DFR process, not the DFR process, because there is no single DFR process.

Not all designers and manufacturers have a DFR process, but even among those that do, their DFR processes are not completely consistent. Many aspects of a company's DFR process are the result of past experiences. In particular, poor experiences and costly experiences lead to practices intended to prevent future failures. Companies that have not experienced similar failures may not emphasize the same elements in their DFR process. Other elements that shape a specific company's DFR process may include the specific personnel they employ and what their employees know how to do, enjoy doing, or have the resources to accomplish.

Over the last few years, many individuals in industry have learned the importance of using structured processes to perform their work. Systems like Business Process Re-engineering, Lean, and Standard Work have shown that a structured approach can add effectiveness and efficiency to almost any pursuit. DFR is no exception. For DFR to be effective, it must be executed in a structured manner. In addition, DFR must be highly integrated concurrently with a similarly well-structured product design process. Simply having a tool box that contains a number of the analytical tools that are used in reliability engineering—and applying them randomly as the design progresses—will not work. Reliability engineers who try to apply DFR in that fashion will find they are never in the right place at the right time. The problems they detect will be identified too late to be addressed. Thus, the project development will move ahead without benefit of effective DFR.

As we discuss the DFR process, keep in mind there are two types of design processes leading to new products. The first type involves completely new products of which there has been nothing similar in the past. In the age of electronics and microprocessors, there are more of this kind of product than in the past. Yet on a relative basis, totally new products still make up a small portion of the total. The second type involves an existing product that is either modified to include new or better features or is being applied to a new application. As you might imagine, there are far more products that fall into this second category than the first.

For both types of design process, the DFR process can be broken down into a series of discrete steps. These steps are organized in a manner that tends to follow along with appropriate phases of the design process. Below, I will describe the System Engineering "V" model, which provides a useful pattern for developing and organizing the information associated with a design. This model is a useful tool for determining the order in which the discrete reliability design steps should be taken. The steps can overlap one another, can be accomplished in a number of ways, and can include a variety of reliability analytical tools. The important thing is that each company understands the objectives of

each discrete step and has a structured approach for accomplishing it in a timely manner.

Organizing the DFR Process

The DFR process must be conducted concurrently with the product design, This section describes a typical product design process as a way introducing the reader to the steps to which DFR must be linked.

The Phases and Order of the Design Process Steps

As mentioned above, the Systems Engineering V-Model, provides a graphic way to portray the various phases of product development.

- ◆ It shows the sequential phases involved in the design and development process.
- ◆ It highlights the fact they are sequential in nature.
- ◆ It clearly shows that the process takes time and that specific elements are performed at specific times during the design process. If the proper steps are not accomplished at the appropriate time, it is impossible to perform them at a later time without going back and repeating a number of earlier tasks.
- ◆ If the customer and owner want to have an impact on choices that are being made, they have to establish requirements or make comments at the right time or miss their opportunity.
- ◆ If the customer and owner want to exercise some amount of oversight, they need to station their inspectors in the right place at the right time, as portrayed by the V-model.

Not everyone's design process is the same. As a result, the DFR processes that overlay on the design processes cannot be the same. In some cases, the design process does not even seem logical or rational from the customer's perspective. For instance, the time needed to gather information concerning the buyer's requirements and to integrate them into the final design may not even exist in the seller's design process. This malady seems to be a fairly common one.

A solution to the malady—one which simultaneously addresses the needs of DFR—is the use of a tool like the System Engineering V-Model shown in Figure 2.1. When using the Systems Engineering V-Model, it is possible to identify both what steps will be taken in the design process and nominally when they will occur. Once this information is known, it will be possible to select the appropriate DFR tool to be used at each point in time to ensure all aspects that affect reliability are addressed in the appropriate manner. It will also be possible to ensure that all parties identify requirements and provide inputs at the right time during the design process.

As mentioned earlier, different organizations may apply different techniques during their DFR process. One point is consistent among all DFR processes: If an appropriate reliability tool is not applied at the right time during the design process, reliability will not be addressed during the design. The syndrome of "Too little, Too late" will again raise its ugly head.

Events shown on the model tend to overlap with the next earlier step and the next later one. But all steps are intended to represent activities that are always moving forward (left to right) on the time line. No step on the right hand side overlaps with any step on the left hand side. In other words, if a critical issue

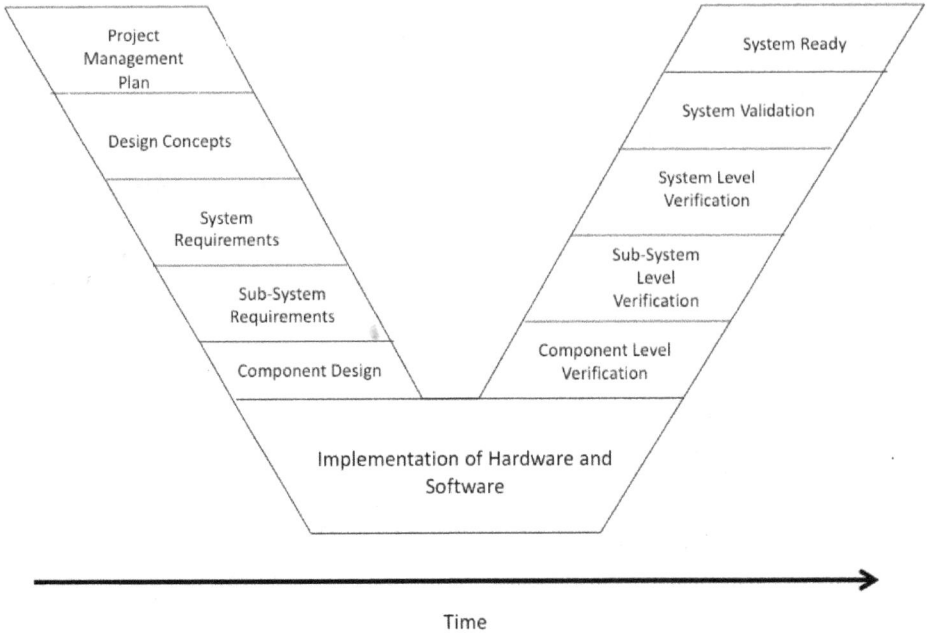

Figure 2.1 The System Engineering V-Model

is missed during the development process, it is impossible to address it properly during the integration and testing phase. Introducing new system requirements while steps in the right hand half of the system-V are underway will necessarily result in significant delays and added cost.

Beyond providing a useful tool for describing the design schedule, the V-model provides a useful tool for communicating precisely how and when the design process is accomplished. For instance, reliability is a characteristic that is largely determined at the component level. If customers and owners want to see how choices concerning components are being made or if they want to communicate specific desires pertaining to specific components, they need to do so before or during the component design phase. Doing so later will either be ignored or will result in added costs for rework.

As we discuss the various techniques applied during the DFR process, keep this V-model in mind. It will be helpful in determining which methods are most appropriate during each step of the design process.

I'm not attempting to pre-empt the sellers' and owners' choices on how best to organize their individual designs and DFR processes. However, it is useful to describe the path of the V-model and the elements of the DFR process that will accompany each phase.

PROJECT MANAGEMENT SCHEDULE

The first step in the process is to publish and agree upon a Project Management Schedule. This schedule identifies when project development will begin, when key steps will occur, and when the project will finish. The Project Management Schedule must be realistic in terms of available resources and the time needed to complete each task. It must also show that the product is delivered on time.

At the conclusion of this phase, the overall process for developing the project will be completely developed and communicated to all who will be involved. Although this schedule may be at a high level (lacking small details) it must contain sufficient detail so key stakeholders and individuals will understand when they must act to avoid delays.

The completed Project Design Schedule can take any of a number of forms including a Gantt chart, a PERT (Project Evaluation and Review Technique) chart, or even a simple spreadsheet showing specific steps and the dates they will be done. The form is not important so long as they are clear, accurate, and communicated the right people in a timely manner.

DESIGN CONCEPTS

The next step in the process follows closely with the Project Management Schedule, but adds greater detail. This step identifies basic concepts that are determining factors of the design. Although customers have the ability to identify a number of specifications, there are frequently more factors that are beyond their control. When buying a product that is simply a new version of an existing model, most choices were made in the past. The sellers can highlight which elements are open to change and which are not.

Another example occurs when the product includes some characteristics that come under regulatory controls, for instance the requirement for pollution controls on new vehicles. These pollution control devices on exhaust systems are required despite the fact that they reduce horsepower and responsiveness and add cost to the vehicle.

At the conclusion of this phase, all participants should understand the key elements of the design process and, specifically, what issues will determine design decisions. In the case of an element that will come under new environmental restrictions that take effect before the new asset is placed in service, this restriction will be a controlling concept associated with the design of that asset.

Suppose a number of choices are available for some design element, but each comes with specific plusses and minuses. It will be important to understand the alternatives and issues that will govern the final choice. This phase may be the first at which some of the reliability design tools come into play. For instance, the Physics of Failure (POF) may actually limit the possible choices—or, at least, individuals may believe that POF restrict their choices. It would be helpful to acknowledge the perceived limitations as early in the design process as possible, Then if they can and should be addressed, changes in mindsets will be made early in the process.

Continuing with the example of a pollution control device, these devices typically operate at extremely high temperatures. In turn, the high temperatures accelerate deterioration due to corrosion. It might be possible to reduce the deterioration and increase the useful life by using an exotic metal in the design, but that step would add to the cost. If the sellers would like to leave the choice of using an exotic metal pollution control device up to the customers, this would be the time to raise both the overall issue and the alternatives.

SYSTEM REQUIREMENTS

The next step is one that is specific to the customers, their needs, and how the product will be applied. In as thorough a manner as is possible, the overall System Requirements must be clearly defined and communicated. Although many obvious requirements—such as capacity, functional capability, and maximum cost—are typically well understood, many other system requirements, —like reliability, availability, maintainability, and useful life—are far more subtle.

Right from the start, it is important to identify the specific requirements for the finished product and to understand how those requirements will create differences between the DFR for sellers and for owners. When determining the reliability performance requirements, the owner is interested in the entire usable life of the asset, thirty years in some cases. The seller is primarily concerned with the warranty period. When quantifying the impact of a failure, the owner will lose a profit opportunity due to lost production. The owner will need to pay for the cost of repairs. The owner may incur costs associated with injuries to employees or public and environment insults. Generally speaking, the seller will only be responsible for costs as limited by the extent of the warranty or significant damages that can be clearly proven in court.

Although the differences between the seller's objectives and the owner's objectives are clearly apparent, it is critical that these differences not be allowed to exist at the conclusion of contract negotiations. The seller must clearly understand the owner's requirements and agree to meet them. At the same time, the owner must understand what the seller is planning to provide and be willing to accommodate the shortcomings. Entering into a contract without either agreement or acknowledged and accepted differences is a mistake.

Allowing unrecognized or poorly understood differences to exist at the end of the design process is even less acceptable. Contract documents must clearly identify the information that the seller must provide to the owner as the design process progresses before construction is allowed to begin. These documents must provide the supporting analysis that clearly identifies:

- The expected reliability during each period in the life of the asset.
- The expected availability during each period in the life of the asset.
- The performance for each key component that plays a part in determining the first two characteristics.
- The frequency, amount, and anticipated cost of any form of

maintenance that is needed to preserve the characteristics described above.

◆ The specific tasks included in the last item and the analysis showing those tasks are maintainable (e.g., Can be done in a manner that restores inherent reliability in a ratable period of time).

SUB-SYSTEM REQUIREMENTS

Sub-systems are groupings of components that exist between systems and components. They are important because they ultimately become the focus of much of the owners' attention. Think of the drive system in a typical vehicle. It is made up of an engine, transmission, and differential. When there is a problem, it typically involves one of the sub-systems. Typically the composite reliability of a system is the product of the reliability of the major sub-systems within that system.

At the conclusion of this phase, requirements identified by the owners must be clearly established. Most operations and maintenance are accomplished at the sub-system level. As a result, for these activities to be consistent with how they are currently handled, the owners need to clearly identify any issues that are important to the integration of the new asset with current assets.

Consider the design of the diagnostic capabilities of a fleet of vehicles. An entire system for performing and interpreting diagnostic downloads on an existing fleet was put in place. The diagnostic system provided information with new units that were not consistent with the existing system; therefore, the new units could not be properly maintained without dramatic changes. It is possible that applying several reliability design tools might have brought this issue to light—the most obvious would be the Living Program. If the sellers were aware of how the owners intended to identify and correct failures and track failure data, it would have been clear that the revised strategy did not fit the owners' requirements.

COMPONENT DESIGN

Components are the smallest element of the device. Failures are the result of deterioration to a specific component. Many choices affecting reliability are decisions that balance the robustness of a costly component with the unreliability of an inexpensive component. As discussed earlier, sellers frequently select components that survive the warranty period. If customers want the sellers to provide more robust components, they must be willing to pay a premium. The customers must make that point clear early in the design process, then enforce it throughout the design to ensure robust components are selected.

At the conclusion of this step, the overall design process will be nearing completion. Design at the component level will determine the need for redundancy. The choice of specific components will also identify the specific Failure Modes common to the components that are selected. Understanding the specific role that components play in the system where they are installed provides the opportunity to perform several forms of reliability analysis. POF analysis can be completed to identify the Failure Mechanisms leading to common Failure Modes. Design choices can then be made to preclude those Failure Mechanisms. One or more of several forms of Failure Modes and Effects Analysis (FMEA) can be completed to ensure that the component robustness or use of redundancy is consistent with the risk associated with the effects of the failure of each component. This is also a point in the design process when it is possible to construct Reliability Block Diagrams (RBD), perform Reliability Centered Maintenance (RCM), and perform RAM analysis to begin the understanding if the anticipated system will meet owner's requirements for reliability, availability, and Total Cost of Ownership.

IMPLEMENTATION OF HARDWARE AND SOFTWARE

As the graphic portrayal of the V-model shows, this step is the transformation from thoughts and words into a tangible product. Before this step, requirements are being developed and after it the success of transforming concepts into steel and concrete are being verified.

At the conclusion of this phase, the design and modeling of the design should be complete. This phase should signal the transition from a paper system to a physical system. It provides the first opportunity to see if the tangible product will behave as expected or if the design was flawed. It also provides the first opportunity to apply reliability elements from the design to the manufacturing process. It may be that the elements that have been assumed in the design process are no longer available. They may not provide the appropriate form, fit, and function as was assumed. The range of capabilities of components may stretch components beyond the range assumed in the design. For instance, when automatic engine starting and stopping systems were installed on locomotives to save fuel, the starters were cycled far more frequently than assumed in the original design. As a result, starters failed after one year of use rather than the expected three-year life.

The initial implementation of hardware and software provides the initial opportunity to see the design in a tangible form that either will or will not perform. In many cases, this step is not performed directly by the seller. Instead, sellers may need to convince sub-tier suppliers to create mock-ups or they may need to ask the owner to perform tests on existing systems to prove or disprove concepts. Through this phase, it is important to remember that the objective is not to get the system to work once. The objective is to create a system that will function for the long haul in any operating conditions including the most severe.

COMPONENT LEVEL VERIFICATION

In the development of a new product or a revision to an existing product, the capabilities of most components are taken for granted. For instance, one would not test Grade 7 studs to verify their strength. On the other hand, there may be specific components upon which the reliability of the overall product rests. There may be components for which the customer has elected to pay a premium to get performance better than that available from the basic component. In these cases, additional steps will be taken to ensure the components deliver the required performance.

At the conclusion of this phase, tests will be completed that have focused on the component level. For instance, individual components can be exposed to extremes in temperature or vibration either for extended periods or using accelerated testing methods. The tests will determine if the individual components in critical services are likely to survive the exposure anticipated by the design and requirements.

SUB-SYSTEM LEVEL VERIFICATION

Sub-systems or combinations of components between the component and system level may be the level that controls performance. In so far as performance is controlled at the sub-system level, the characteristics being controlled at that level should be tested and verified. An example from my own experience involves burners installed in a process furnace in a refinery. The burners were new, quiet, ultra-low-NOX burners. Unfortunately they created too great a restriction for combustion air causing the furnace to "surge" or seemingly gasp for air. Not long after startup, this resulted in a flame-out. When unburned hydrocarbon reached a hot surface, it exploded, causing severe damage to the furnace shutting down the unit. Adequate testing and verification had not been completed.

At the conclusion of this phase, tests will be conducted focusing on the sub-system level. The objective of this testing is to identify issues resulting from interaction of components or between components and software at the sub-system level. Frequently, components behave as expected alone. But when asked to perform specific functions or to do so in response to specific computer instructions, they behave in an unexpected manner. By performing tests when applied in a sub-system, it is possible to identify many problems while avoiding the complexity of larger systems.

SYSTEM LEVEL VERIFICATION

Typically, complete systems perform the functions for which a product has been purchased. The ability to perform complete functions cannot be accomplished at any lower level, so it is important that key functions be verified once complete systems have been assembled. Keep in mind that failures are actually instances in which the product is unable to perform its intended function. Therefore, it is critical that systems be tested and verified for their ability to perform at required reliability and availability.

At the conclusion of this phase, tests will be conducted focusing on the overall system level. Once subsystems have been shown to perform as needed, it is possible to assemble the sub-systems into complete systems. If sub-system specific problems have been identified and addressed at that level, it can be assumed that system level problems are the result of unanticipated interactions between subsystems. Knowing this helps focus corrective action on interactive elements (like interactive software instructions) rather than the entire range of possible problems.

SYSTEM VALIDATION

Frequently, products consist of more than one independent system. In many of these cases, the functioning of one system can have an unexpected impact on one of the other systems. As a result, it is important that all systems that are part of a product be combined and allowed to operate in a manner that exhibits all possible failure modes. For example, a system may create a resonant frequency vibration that produces deterioration; yet the deterioration would not be evident if the system were operating separately. In this case, testing and verification in combination are critical.

At the conclusion of this phase, the overall system should be shown to meet all requirements and expectations. Although earlier steps of verification has been made by comparing performance to specific requirements and specifications, this step challenges the system to perform all on-going requirements when operating in the context of rea-world requirements. At this point, the system is exposed to all the circumstances driving failure that exist in the actual operating environment, including those the designers and owners may never have considered.

SYSTEM READY

At the conclusion of this phase, the asset is in service, performing the intended functions. The remaining reliability tool to introduce is the Living Program. In addition, it is important for the owner to continue to look back on all forms of earlier analysis to see that all elements are performing as expected and required.

Let's now return to the typical elements of a DFR process, describing each in greater detail. It should be clear to the reader from the discussion above where each of these elements fits with the concurrent Design and DFR processes.

Integrating DRF with the Design Process

You now understand the phases and nominal timing that the sellers will use in the design of your new asset. Assume you want to take the steps needed to ensure that reliability is being adequately addressed during those phases. It will now be necessary to select from the number of reliability design tools that can adequately meet your objectives. As you will see in the following sections, in some cases, several different methods can be used to perform various forms of reliability analysis. Within the various methods, different techniques (e.g., streamlined

RCM or classical RCM) can be chosen as well as software packages to support these techniques. Software packages differ in the way they perform the analysis or display the results.

Although it is important to select the right tool at the right time, two aspects of the selection are even more important:

1. Sellers and owners must be consistent in their choice of tools and methods. If they choose different methods for testing or verification, or use different software for analysis, unnecessary disagreements will result from differences.
2. The reports being produced by the methods and tools presented below are ultimately intended to facilitate communication of relatively complex issues. Owners and sellers must both understand what is being said as well as what is meant. Failure to achieve such clear communication may significantly reduce the value of the entire DFR process.

We now continue this discussion with a description of the activities needed to address reliability aspects during the design process.

Determine the Reliability Performance Requirements.

The process of identifying and communicating performance requirements is the first and most important step in the DFR process. Although there may not be any special methodology or supporting software associated with this step, it is nonetheless a fairly sophisticated activity. An example of the difficulty that might be experienced is provided by reviewing the various ways that reliability of components can be stated. For instance, locomotive reliability is typically stated in one of two ways—FLY (Failures per Locomotive Year) or MMBF (Mean Miles Between Failure). As the number of miles a locomotive is used increases, the MMBF will increase, but the FLY may decrease. Because locomotives are serviced and maintained on relatively frequent intervals, it is very important that they are capable of surviving between contacts. Therefore, it is also possible to consider Mission Reliability. Generally speaking, locomotives are made up of a myriad of components, none of which have their basic reliability capabilities stated by manufacturers in terms of any of these measures.

In this case, it is important to select a common measure that is meaningful to the circumstance in which the asset is being used. Then all other measures of performance must be concerted to a form that supports the overall measure.

There are a plethora of reliability performance requirements that can be identified. As mentioned above, the choice of performance measures and the selected level of performance are areas where the requirements set to meet the owner's needs can depart from those needed to support the seller's objectives.

FAILURE FREQUENCY / RELIABILITY

The first requirement is the failure frequency or reliability. Ultimately both the sellers and the long-term owners are interested in how frequently the overall asset fails. As a result, they are also interested in the frequency at which the components used to construct the system fail. The basic definition of the term *reliability* is the frequency at which an item will fail, or be unable to perform its intended function over a given period of time. For instance, a specific reliability may allow for a single failure in a one-year period. As with any highly technical definition, it is possible for a requirement that seems simple and straightforward to be interpreted in a number of ways. For instance, the definition of a failure can vary. I recall one interpretation that did not count failures that entailed a loss of function that lasted less than four hours.

It is also important to understand how long the specified reliability is intended to last. A seller may guarantee the reliability for only the warranty period. The owner may expect that, with proper operation and maintenance, the specified reliability will last the entire life of the asset. There can also be differences in terms of what is viewed as a failure. The seller may view failures as situations when the asset is "down and dead." The owner may view failures as any situations in which the asset fails to perform the intended function to the fullest extend of the specifications. Consider a 4400-horsepower locomotive that is pulling at only 4200 horsepower. Is that performance a failure or not?

That question may be answered in one manner by one part of the owner's organization and in another way by another part of the organization. In the case of the locomotive operating with reduced horsepower, the following perspectives may exist:

♦ Even at reduced horsepower, the locomotive will finish its mission; in that respect, it is reliable.
♦ The locomotive is not able to pull the amount of freight for which it was purchased or at the speed it was intended to operate (leading to rail system congestion); the event is a failure.
♦ The entire consist (group of locomotives pulling the train) will operate in a less efficient manner with one unit operat-

ing at partial efficiency; the event is a failure in terms of efficiency.

♦ Allowing this condition to go on uncorrected is unacceptable and, therefore, the unit must be sent to the shop and repaired. The shopping will result in a period of unavailability; this event is viewed as a failure.

It is best to define reliability clearly both in terms of the frequency of failures and the cost of those failures during each year of the usable life of the asset. (The costs include all forms of out-of-pocket expenses as well as lost revenues when the asset is not performing its intended function.) We must also avoid confusion over what is meant by the term *failure* and how long the specified reliability should last.

AVAILABILITY

A second performance requirement that should be set is availability. Although the term *reliability* refers only to unplanned events that result in loss of functionality, availability is affected by both the planned and unplanned outage events. Availability is defined as the portion of the total time an asset is able to perform its intended function. An asset is unable to perform its intended function during the time it is down as the result of poor reliability. In addition, an asset is unable to perform its intended function during the time it must be shut down for planned and scheduled maintenance. The periods of planned maintenance can include overhauls, outages, or turnarounds (major renewal events) as well as not-so-major events that are necessitated by the limited usable life or one or more of its components.

Availability is described as:

$$\text{Availability} = \text{Uptime} / \text{Total Time}$$

Said another way, availability is the time an asset is able to perform its intended function divided by the total time possible. For instance, the time that a plant has been mothballed or an equipment item like a locomotive is stored is not part of the total time possible.

Another way to determine availability is to use the statistical means of the times between failure and the times to repair.

$$\text{Availability} = \text{MTBF} / (\text{MTBF} + \text{MTTR})$$

where

MTBF = Mean Time Between Failure
MTTR = Mean Time To Repair

As with reliability, the seller and the owner may have different viewpoints on the calculation of availability. One significant difference may be the time at which the major renewal event is required. A good example is an overhaul for an engine-driven asset. If the usable life is assumed to be thirty years and the overhaul period is seven or eight years, the asset will require three overhauls during its life. On the other hand, if the overhaul interval is more than ten years, the asset will require only two overhauls. Because overhauls can cost a significant portion of the initial cost of the asset, the difference between two and three overhauls can be significant when calculating the Total Cost of Ownership (TCO) or Life Cycle Cost (LCC).

Again the best way to be clear on this subject is to ask that costs over the entire life cycle be described on a simple bar graph (see Figure 2.2) or in a spreadsheet. This will avoid con-

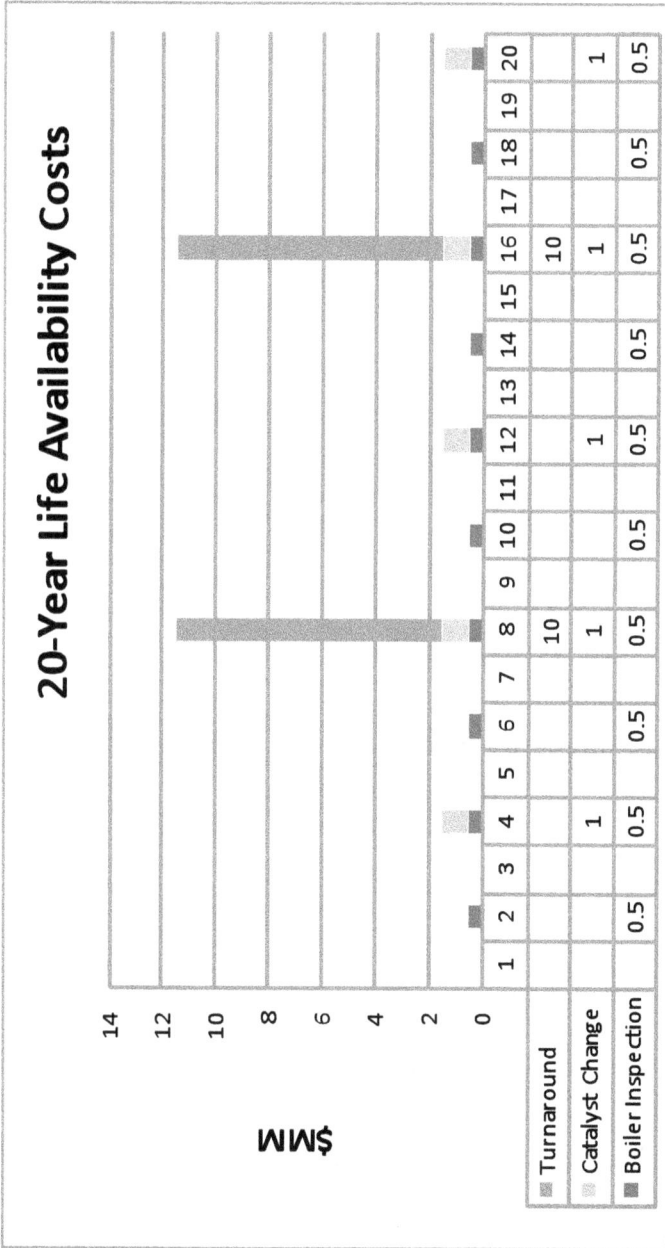

20-Year Life Availability Costs

$MM

	1	2	3	4	5	6	7	8	9	10	11	12	13	14	15	16	17	18	19	20
Turnaround								10								10				
Catalyst Change				1				1				1				1				1
Boiler Inspection		0.5		0.5		0.5		0.5		0.5		0.5		0.5		0.5		0.5		0.5

Figure 2.2   20-Year Life Availability Costs

fusion concerning:

1. The timing of events
2. The cost of events
3. The length of the usable life

Although relatively simple, Figure 2.2 provides future owners with a clear idea of when outages affecting availability will occur and nominally what they will cost. The objective is to avoid distasteful and unexpected experiences.

MAINTAINABILITY

A third performance requirement that should be determined before the development of a design for a new or modified asset is maintainability. Maintainability is a concept more abstract than either reliability or availability. As a result, it is more difficult to quantify and more difficult to produce agreement between the seller and the owner of an asset.

Maintainability is a measure of the ability to return the asset to its full inherent reliability in a ratable period of time. Inherent reliability is the maximum reliability possible from an asset; it is a measure resulting from configuration and component selection. If an asset is operated as well as possible and maintained as well as possible, it will be possible for it to achieve its full inherent reliability. The full inherent reliability may best be described as a condition that will be achieved when an asset is nearly new, but past the point that any infantile failures are possible.

An example I have used numerous times is the situation when you take your car in for repair. If the mechanic says, "I don't know how long it will take, but it will be good as new when I finish," the car is not maintainable because the repair cannot be done in a ratable manner. If the mechanic says, "I will have it done in two hours, but I don't know how long it will last," it is not maintainable because there is no assurance it will be returned to the full inherent reliability. It is necessary for the mechanic to say, "I can fix it in two hours and it will be as reliable as new" for the car to be truly maintainable.

In order for sellers to be able to speak knowledgeably about the maintainability of an asset, they will need to know what maintenance will need to be performed on the asset over its usable life. Here we are talking about both proactive maintenance (predictive and preventive maintenance) and reactive maintenance (repairs). In order to have a thorough understanding of the proactive and reactive maintenance that will be required, it will be necessary for the sellers to understand both what is likely to fail and what activities must be completed both to intervene before failure and to ensure the longest life of components that make up the asset. Once the sellers understand the tasks that must be completed, they will need to perform simulations or "dry-runs" of the tasks to see if they can be done in a ratable manner and if those tasks are certain to restore the inherent reliability.

In order to identify all the tasks, it will be necessary to perform some kind of analysis like Reliability Centered Maintenance (RCM). This analysis should identify both the likely failures and the predictive and preventive maintenance needed to maximize the interval between those failures. Once the proactive and reactive tasks are identified, it will be necessary to perform enough analysis to clearly understand the key elements of the tasks. For example, is there ready access? Will any craft persons be able to quickly find the failed component or will they have to hunt and guess? Will unsure repair steps be needed, like allowing a drying period for an adhesive? Once the repair is complete, is the reliability of the repaired device assured or is there a reasonable chance it will fail again on testing or soon after release?

Maintainability and "maintainability reviews" are much more clearly defined than they were twenty years ago. Back then, they involved simply walking your tallest craftsmen through the asset to see if they bumped their heads in areas where headroom was limited. A good example of a highly maintainable asset is a submarine in which radio modules are quickly removed and replaced with "hot-running" spares when active components show early signs of failure.

LIFE CYCLE COST (LCC)

The final area of reliability performance requirements that should be determined comes under several titles: Total Cost of Ownership (TCO), Life Cycle Cost (LCC), or Cost of Un-Reliability (CoUR). For simplicity, we will use the term LCC for this discussion.

Although some aspects of Life Cycle Cost may seem somewhat redundant with several of the areas already described, LCC is different in its level of completeness; it is worth any repetition that may occur. This area is the one in which the differences between the owners' business model and the sellers' business model will become most pronounced. Ultimately, total LCC will be a result of the overall reliability and robustness. LCC will be shared between the owner and the seller. Generally, sellers are accountable for many costs during the warranty period and costs

associated with the asset's inability to fulfill guaranteed performance (as can be proven by the owner). The remaining costs are
the responsibility of the owners.

Owners want to see that all performance requirements are
clearly defined along with the sellers' accountability for situations when those requirements are not met. Again, a description of the expected total costs is an item that is best described
using a simple bar graph (Figure 2.3) or spreadsheet that clearly describes the expected costs for each year of the asset's life
as well as who is accountable for the costs. If the costs are different than are expected by the owners, then the reliability,
availability, or maintainability will also be different.
"Dollarizing" the value of each of those characteristics is a
good way to ensure that everyone is clear on expectations and
requirements.

The LCC of an asset has the following elements:

◆ *Initial cost*
◆ *Operating costs*
  • Cost of operating personnel
  • Fuel costs
  • Cost of operating materials
◆ *Maintenance costs*
  • Cost of labor
  • Cost of material
◆ *Costs related to failures*
  • Value of lost production
  • Cost of repairs
◆ *Costs related to planned outages or overhauls*
  • Value of lost production
  • Cost of maintenance

♦ Costs related to untoward events that result from failures
  • *Cost of safety incidents*
  • *Cost of environmental incidents*

A chart like the one in Figure 2.3 on the following page is useful because it shows whether the cumulative costs over the life of an asset are equal or greater than the initial cost. If they are, choices made on the bases of initial cost alone will short-sighted. Understanding the cost of ownership over the entire life cycle provides owners with information to make farsighted decisions.

As either an owner or a seller, your business model is based on certain assumptions concerning levels of production, income, and costs. Rather than trying to deal with both sides of the ledger, reliability engineers typically deal only with the cost side of the ledger. Their focus is not about profit potential, but instead the cost of lost opportunity when the asset is unable to produce. This area typically separates owners from sellers. Sellers might say that the owners' lost production is important to them, but their concern typically does not translate into real dollars for the sellers as it does for the owners. The owners must somehow consider this cost in vendor selection. In doing so, they ensure that the sellers who are ultimately selected to pro-vide the asset also consider lost profit opportunity in their design process.

For completeness, a complete list of typical expense cate-gories, along with a detailed description, is included in Appendix 5 at the end of the book.

Total Costs Over 20-Year Life

- Initial Cost
- Operating Cost
- Maintenance Costs
- Cost of Failures
- Overhaul Cost
- Cost of Liabilities
- Cumulative Post Start Costs

	1	2	3	4	5	6	7	8	9	10	11	12	13	14	15	16	17	18	19	20
Cumulative Post Start Costs	45	90	135	180	225	290	335	380	425	470	515	580	625	670	715	760	805	870	915	960
Cost of Liabilities										100										100
Overhaul Cost						20						20						20		
Cost of Failures																				
Maintenance Costs	25	25	25	25	25	25	25	25	25	25	25	25	25	25	25	25	25	25	25	25
Operating Cost	20	20	20	20	20	20	20	20	20	20	20	20	20	20	20	20	20	20	20	20
Initial Cost	500																			

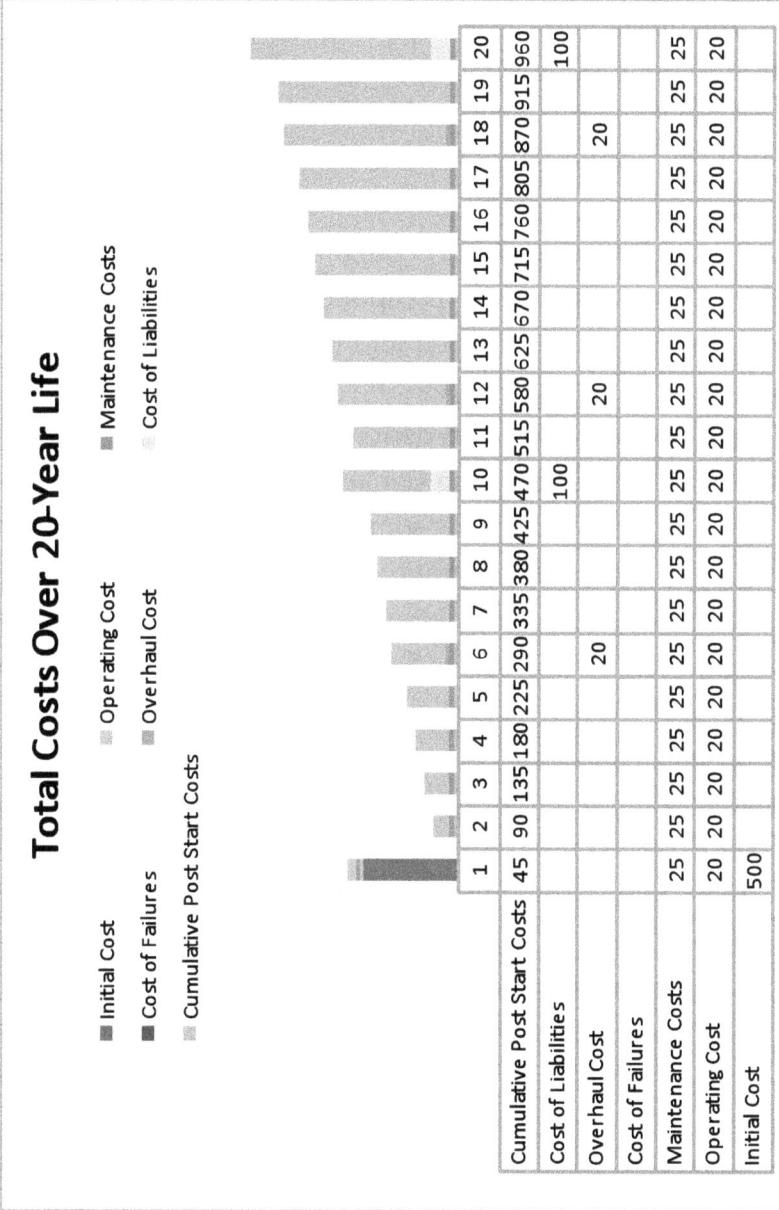

Figure 2.3 Total Costs Over 20-Year Life

Change Analysis: Identify Changes from Already Known Conditions in Service or Application

In this discussion, we will focus primarily on the myriad products that are updates of earlier products or adaptations of existing products into new and different applications. Rather than focusing exclusively on the changes, we will look more comprehensively at opportunities for improved reliability as the next generation of a product is being prepared for release.

The various elements or components of a current generation of an asset can be grouped into the following categories:

♦ Current elements for which performance is adequate and the element is viewed as being stable:

This category describes some subsystem or component within the proposed generation of the asset for which performance fills all the requirements and nothing is going to change. Both the design and the loading on the subsystem will remain the same. Furthermore, all suppliers of components and specifications of components will remain constant.

♦ Current elements for which performance is not adequate:

This category describes those elements of a system that are neither changing service nor changing source, but have not been providing the needed performance. There should be lots of anecdotal data available on these elements, but that data is typically not well organized or well defined. Had it been so, any problems would have already been solved.

♦ Current elements for which performance is adequate but are not viewed as stable:

This category describes the category of unexpected prob-

*lems that is probably most common. All too often, the com-
panies producing systems tend to take the reasons for their
good luck for granted. Good luck is often the result of a lot
of hard work, structure, and discipline on the part of a sec-
ond- or third-tier supplier. Changes in sources of components or in the manufacturing or inspection requirements
can produce unexpected poor results. Any and all instances
in which changes are being made for any reason should be
placed into this category and addressed in the Reliability
Block Diagram (RBD) analysis as a source of risk.*

♦ Altered elements for which there is a sufficient amount of
usable experience:

*This category includes items for which a change has hap-
pened, but the change is one with which the manufacturer
is very familiar. Suppose a seller starts doing business with a
new owner, but the new customer is in the same business as
one or more current owners. In this case, it might be
assumed that the new customer will use the asset in the
same manner as the other current customers. That is a reasonable assumption on the surface, but would require some
amount of verification. Experienced companies may pride
themselves as having a competitive edge by being better
operators than their competition. In some cases, that source
of pride is real; they spend more time developing their personnel and, as a result, have far better results.*

♦ Altered elements for which there is not a sufficient amount
of usable experience:

*In some cases, an owner plans to use an asset for a serv-
ice that is entirely different than it was used in the past.
Generally speaking, there is no reason why the asset would*

not work in the new service, but it has never been tried. In this case, there is a great deal of value in testing the system against the stresses and environment of the new service before assuming everything will be OK.

Think of how many improper uses you have found for a standard screwdriver (chisel, pry-bar, scraper, etc.). Now recall where it worked satisfactorily and where it failed. Hammering a plastic handle typically ruined the tool.

Identifying the elements that fit into each of these categories and properly sorting them will depend on

♦ Information Systems that **thoroughly track the actual performance of components currently in service.**
♦ Manufacturing Information Systems that identify changes in suppliers or changes being made by current suppliers; these changes will result in differences in the robustness of their products.

Once Change Analysis has identified the situations where a change will occur, several reliability tools can be used to assess the impact of the change. Physics of Failure (POF) analysis can be used to help understand the forms working on the changed elements. (See the next section for greater detail.) Understanding the POF that exists in the environment where the product will operate is the first step in identifying the possible Failure Mode or Modes that must be considered.

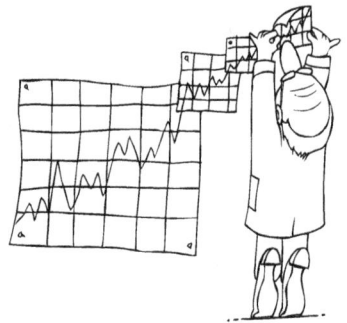

The tool commonly used next during this stage of the DFR process is FMEA—Failure Modes and Effects Analysis. Change Analysis and POF identify the Failure Modes that must be considered. Understanding the effects produced by the failure of similar components in similar situations is useful in identifying the effects that must be considered in the case being studied. Tracking the current Failure Modes and the Effects produced as a result of those Failure Modes is more valuable and more valid than all the analysis produced by perceptions and assumptions. Furthermore, changes in systems and changes in applications can produce Failure Modes. These must be added to the current FMEAs to understand the impact of changes and to identify corrective actions.

Other tools are Risk Analysis and Risk Reduction. Risk Analysis can be either integrated with FMEA or conducted separately. Most computerized forms of Reliability Centered Maintenance (RCM)—which is a form of FMEA—typically include some automated form of risk analysis that facilitates organized risk reduction.

Risk equals Impact times Likelihood. In turn, likelihood can be attached to each Failure Mode and Impact can be associated with each Effect. Therefore, it is possible to use the information generated during a FMEA to quantify the current risk. Also, the results of FMEA can be used to prioritize the value of various risks so the areas of greatest risk can be addressed first. Finally, it is possible to quantify the reduced risk after some change has been made (e.g., the change of a component, the addition of redundancy, or the addition of proactive maintenance). Subtracting the revised risk from the current risk, it is possible to quantify the risk reduction.

The data coming from both FMEA and RCM can be portrayed in a spreadsheet such as the one seen in Figure 2.4. The spreadsheet shows the hypothetical event (a Failure Mode), the likelihood of that event, and the impact of that event. (RCM is simply a form of FMEA that specifically focuses on reliability and the maintenance needed to ensure a specific level of reliability.) Typically the Failure Modes being described are time dependent; they are more likely to occur as time passes. Therefore, it is important to describe the time dependency, if one exists. Frequently all of these factors are summarized in a single number that allows the total value of the risk for each failure mode to be compared. If there are time dependencies, the risk may change over time. As a result, the spreadsheet may also describe the various risks for various time periods.

A variety of names are used for the one number that quantifies the risk associated with each failure mode. Many commercial software tools can perform the analysis and store the results. Subtle differences between software and terminology are not important. What is important is that the analysis is accomplished in a timely manner. The results are used to identify which of the risks must be addressed to produce a system with acceptable performance.

Once the risk analysis is complete, the results are typically arranged with the largest number or highest risk first and the other failure modes in decreasing order of importance. In addition, the individual risk codes are added together and a column showing the cumulative risk provided. In situations where the risk changes with time, there are a number of spreadsheets, one for each year in the life of the asset.

Failure Risk Analysis					
Asset Name: Go-Fast Scooter		Period Covered: First Year of Operation			
Number	Failure Mode	Likelihood	Impact	Risk Index	Cumulative Risk Index
1	Brake -Total Failure	0.01	1,000,000	10000	10000
2	Engine Cylinder - Catastrophic Connecting Rod Failure	0.005	250,000	1250	11250
3	Fuel - Fire	0.001	500,000	500	11750
4	Night Light - Connector Failure	0.1	5,000	500	12250
5	Night Light - Bulb Failure	0.1	3,000	300	12550

Figure 2.4 Failure Risk Analysis

This table can be used in two ways:

♦ Sellers typically have some risk index (combined risk factor) above which they will not accept. For failure modes that have a risk index greater than the maximum, the sellers will apply some form of remediation or corrective action to reduce the risk. For instance, if a specific failure mode is the result of a failure mechanism like corrosion, the device subject to deterioration might be better coated or replaced with one of improved metallurgy.

♦ Sellers frequently have some cumulative risk index above which they will not venture. Although a single failure mode might not be overwhelming, the "million bee stings" resulting from a large number of relatively low-risk failure modes may cause inadequate performance. In this case, the sellers may select several failure modes that can be eliminated or mitigated to reduce the overall failure index for the asset. They will look at the cost effectiveness of the possible alternatives. In other works, which failure modes can be addressed most easily with the least cost?

Unfortunately, many sellers deal with these spreadsheets differently during the warranty period or the period over which they are accountable for failures than they do with later periods. For instance if the asset has a five-year warranty, and the seller will accept a risk equivalent to $1,000 per year during the warranty period and $25,000 per year after the warranty period, the owner is in for a distasteful surprise.

The value of risks ultimately becomes a real cost. Owners much understand and address significant changes in the risk of failure as a product ages. If the sellers are using risk analysis to delay failures past the warranty period, the owners must understand the implications. They must take steps needed to ensure the failure risk remains at an acceptable level over the entire life of the asset—or be prepared to pay for risks they have accepted.

Perform Analysis and Quantify Risks

We now continue from the step above and apply many of the same kinds of analysis to all elements of the new asset. The objective of this next step is to complete an analysis that is sufficiently comprehensive to quantify the total risk of failure for the complete asset.

There are several differences between the analysis performed as a part of the first step (change analysis) and this comprehensive step:

♦ The analysis based on change can and should be conducted as soon as the change is recognized because the likelihood of time consuming corrective action is greater. In some cases, the need for change might be identified right at the start. For example, an owner might want an asset that is larger than any produced earlier. This signals a number of changes. On the other hand, changes may come later in the process as the result of inaccurate assumptions. For example, a supplier might have changed its manufacturing process without notifying the seller.

♦ This more comprehensive analysis can be based largely on experience and data coming from current assets with which the seller and owner are familiar. As a result, much of the information and prior analysis can be assembled early in the design process, if it exists. The act of assembling this information will help sellers and owners understand how much of the information needed to perform the analysis currently exists and how much work lies ahead.

If the first step of the analysis has been done in a thorough manner and if the new product is similar to past products and applications, this step will be simplified. On the other hand, if history and experience have less usable information to offer, this step will be more labor intensive.

As with the earlier step of the DFR process, Physics of Failure (POF) is an element of analysis used at this stage of development. POF analysis identifies the physical causes and Failure Mechanisms the asset will experience during its life. For example, mechanical devices experience deterioration from four Failure Mechanisms: Corrosion, Erosion, Fatigue, and Overload. If an asset is to operate in a wet environment, and if the asset is composed of several different kinds of metal, one of the failure modes that should be expected is corrosion. During POF analysis, corrosion is identified as one of the failure modes. The various Failure Modes that can result from this

Failure Mechanism can be identified.

Assume that a critical electrical switch is left unprotected and is exposed to a corrosive environment—the POF analysis has identified the Failure Mechanism. It has also identified the Failure Mode (Switch Failed Open – Corroded). POF analysis will also identify the Effect that will result from that failure. In this case, the Effect may be complete loss of function and a resulting outage of the asset. Depending on the availability of a new switch and the ease of replacement, the duration of the outage can be estimated along with the resulting cost.

As described earlier, Risk equals Impact times Likelihood. POF analysis has identified the Effect and the value of the effect. All that remains to understand the risk of failure is to quantify the likelihood. In this example, the failed switch contained at least two different metals, the environment was wet, and the switch was not protected from the moisture. The question of likelihood was not one of "if," but one of "when." In other words, the situation is set up for certain failure. The only real question is when the failure will occur. Here, the likelihood of failure might be quantified as 25% in the second year, with 50% cumulative likelihood of failure by the second year, and effectively 100% cumulative likelihood of failure by the third year.

In this situation, the analysis has identified a highly likely and costly failure. It would be foolish not to address the situation in some manner. In all likelihood, some form of protection from the moisture would be provided for the switch.

This step of the DFR process is another that might be addressed using some form of FMEA and Figure 2.5. FMEA is valuable because it ensures a thoroughness that would not be provided by a random analysis. Most successful FMEA processes are structured so that each and every element of an asset is analyzed in the broadest array of possible operating conditions and environments. This 360-degree viewpoint guarantees nothing is missed. Clearly POF analysis is useful in understanding the science leading to possible failures. However, it does not provide the structure or discipline to assure a comprehensive analysis. When combined with FMEA, POF provides both a comprehensive and farsighted analysis of all the possible risks of failure.

A simplified approach for applying POF, but doing so in a manner that ensures a comprehensive analysis, is to create another table. If the components being considered are mechanical, then the failure mechanisms are corrosion, erosion, fatigue, or overload. Figure 2.5 shows the table:

POF Spreadsheet						
Equipment Name: Go Fast Scooter						
		Failure Mechanism				
Component Number	Component Name	Corrosion	Erosion	Fatigue	Overload	Source of Deterioration
1	Brake shoe	X				Corrosion-Scuffing after paint is worn
2	Brake Lever Arm			X		Repeated application at levels greater than fatigue limit
3	Seat Springs				X	Use by overweight adults
4	Lighting Circuit	X				Water intrusion into areas of differential metal contact
5	Lighting Circuit			X		Vibration of support bracket
6	Lighting Circuit				X	Over-current when under-voltage is applied

Figure 2.5 POF Spreadsheet

The example above included failure mechanisms associated with mechanical components. It is equally possible to create a similar table for electrical and electronic components. In that case, the failure mechanisms to consider are:

- ◆ Overload
- ◆ Electrical Supply Transient
- ◆ Load Stall
- ◆ Electrical equivalent of fatigue
  - • Persistent loading to levels greater than rated capacit but less than will cause instantaneous breakdown
- ◆ Insulation breakdown
  - • Heat
  - • UV exposure
  - • Chemical Exposure
- ◆ Mechanical failure
  - • Abrasion
  - • Loosening

Ultimately, once the POF analysis is complete, the designer will need to find ways either to interrupt the progress of the failure mechanism or to include the resulting Failure Mode in the FMEA. The first part of this process is a reality check: Is this realistically possible? For instance, overload of the battery circuit is pretty unlikely. The second part of the process is a test of: Do I really care? If the springs break because an overweight adult is riding a scooter made for a child, should the seller care?

For the few items where the POF analysis has identified real and realistic failure modes, those failure modes should be subjected to the forms of analysis described above. The goal is to reduce the likelihood of each real failure mode to an acceptable level.

Take Action Required to Align Performance with Requirements

This is the step in which the designer begins the process of modifying the design to meet the requirements established in earlier steps. Suppose the seller has a product that currently does not meet the defined requirements. Let's also say that that only one component has inadequate reliability. The corrective action process would involve simply identifying a replacement for the inferior component that, when added to the system, will deliver the required performance.

For designers to have found the single inferior part in the example, they would have had to develop a way of testing the system that clearly highlighted the weakness of that component. The process of designing such a test is called Design of Experiment; the objective is to create a situation that will 'turn failures on" and "turn failures off" simply by making some change to the targeted component. Once the designers have isolated the specific component causing the inadequate performance, the next step is to identify a replacement component that will experience the same stresses and strains, and deliver an adequate level of performance. Their experience in developing an experiment that will turn failures on and off at will provides the designers with insight needed to identify a replacement component that will deliver the required performance.

Now think of a highly complex system consisting of many components. All components are behaving in a manner that is consistent with their own inherent reliability, but also in a manner that may or may not support the required overall system performance. The objective of this step becomes one of identifying a number of components that may be inadequate—components that may cause failures in a variety of different situations and environments. In this case, the value of Design of Experiment becomes more apparent. The designer will create situations (extreme cold, extreme heat, extreme humidity, extreme dustiness, any form of stressor) that affect specific components in a manner that will result in a system failure. The objective is to introduce stressors that are within the realm of what should be reasonably expected. As before, the objective is to be able 1) to turn failures on by introducing the stressor and 2) to turn them off by introducing a more robust component while operating at those conditions.

The conclusion of this step of the DFR process is to identify all elements that will experience failures while operating within expected environment and stress levels. Furthermore, the conclusion involves replacing those components with ones that will survive without failures at the extremes of those conditions.

Obviously, this step is much more difficult when dealing with totally new products than ones than are simply adaptations of earlier products. In a totally new product:

♦ There is no experience where sensitivities have existed in the past.

♦ There are no current examples that can be changed to include new modifications.

♦ There are no current examples that can be operated in the new environment.

♦ There are no current examples that can be operated under new loads.

As a result, the first production run of any product is likely to be a test bed and source of a great deal of learning. If a current proven version of the product exists, it is possible to simply start using it in a modified condition, in a more severe environment, or under even more severe stressors. If the test model fails, no harm, no foul. The intent was to see it there were failures, then go through the Design of Experiment process to identify the specific improvements that will be needed to eliminate all failures.

This step in the analysis is at times the most difficult because it is the step that requires the seller to make costly changes to achieve requirements. It is also the step in which many sellers tend to make compromises that undermine the product. When changes become too expensive, some sellers find ways to rationalize either the meaning of the requirement or the results of the risk analysis to achieve a better "fit" with their business model.

For example, if a test device fails during testing, and the replacement component needed to eliminate all failures will add significant cost or reduce the seller's profit, the seller may find ways to rationalize the failures. The seller might choose to believe the test conditions were extreme conditions that were beyond the extremes to which the asset will "normally" be exposed.

There is an old saying that there is nothing less fortunate than the engineer who designs levies to handle 99-year floods the year before the 100-year rains. Few sellers will design their products for 100-year conditions. The greatest likelihood is that the asset being sold will survive the three-year warranty period and even the 30-year life if built to the lesser standard.

Computerized DRF analysis uses Monte Carlo simulation to portray the statistical failures. It does so over not one but hundreds of simulated life cycles. As a result, the analysis will identify all the possible failure scenarios. Even if the simulated life cycle is only 30 years long, it is very likely that the 100-year rains mentioned above will occur during one or more of those simulations. The extreme results will provide the designer with a clear indication of what is possible and what needs to be addressed.

In the case of a complex asset that fails because of a statistically unlikely event, the asset will look the same as one that fails in the same timeframe, but as a result of far more marginal design and acceptance of greater risk. The only way for owners to know that the asset was built to an acceptable standard is to be provided with more detailed information concerning the reliability of components. We will discuss this more in the next chapter covering DFR for the owner's needs.

Verify the Adequacy of Your Design

Once the asset design is complete—including any elements of DFR that are done—the asset should be exposed to detailed product testing. One might ask if this testing is different from the testing described in the previous step of the process. The primary difference is that this testing is expected to address the performance of a design-complete product. All the detailed flaws should have been addressed entirely; failures should occur only at the expected or specified rate. Testing in the last step was needed to determine what is reasonable to expect.

As a result, this stage of testing is intended to simulate the overall life of the asset. To do so, the testing must be accelerated in some manner to represent the same amount of stress the device would see over its entire life. Depending on the quantity of a product that will ultimately be produced, it is necessary to calculate either the number of demonstration units that must be tested or the duration of testing, In either case, the amount should be enough to produce statistically representative results. For products that are produced in extremely large quantities, it is necessary to produce and test large numbers of demonstration units. For products made in smaller numbers, the number of demonstration units can be smaller.

Much of the logic behind this requirement has to do with manufacturing variation. For large numbers, it is believed that the variation in both the components that go into the product and the manufacturing steps used to assemble those components probably have greater variation than in cases where only a few are made. This variation may or may not be true, depending on the care that is exerted in each case. For example, in dealing with the creation of a single plant, variation is avoided at its source by application of quality control to each and every critical element.

Recently I heard a TV commercial for a specific brand and model of car. The commercial emphasized the car's reliability by saying that more than a million miles had been driven during tests. If the test fleet contained ten cars; each test vehicle would have 100,000 miles—assuming all cars were tested the same distance. If the test fleet contained twenty vehicles, each would have 50,000 miles. If there were forty vehicles in the test fleet, each vehicle would have been tested for 25,000 miles and so on. Although the commercial sounded impressive, the actual test being described did not necessarily represent the requirements of the typical car owner.

Independent of how the seller may want to portray the testing, there are several forms of accelerated testing applied during new product development:

HALT—*HIGHLY ACCELERATED LIFE TESTING*

♦ HALT is used during the DFR process while the system is being designed and prior to full-scale production. During HALT, systems are exposed to high stresses at a frequency rate much greater than would be expected during their lifetime. Exposure to extreme conditions is done to force failures to occur so failure mechanisms can be identified and corrected.

HASS — *HIGHLY ACCELERATED STRESS SCREEN*

♦ Like HALT, HASS exposes the product to situations of extreme stress. The difference is that HASS applies the stress after the product is complete. Here the philosophy is that if the product survives the test, it will survive normal use. An example of this kind of testing frequently applied to electrical systems is a Hy-Pot during which a high electrical potential is applied. If the insulation fails, the system fails the test and is discarded or remanufactured.

HASA—HIGHLY ACCELERATED STRESS AUDIT

♦ HASA is much the same as HASS, except it is applied only to a representative sample of the product. If any portion of the selected sample fails, the sample size will be increased until the auditor is satisfied all defects have been identified. With HASA, it is important to employ the services of someone sufficiently familiar with statistics to determine the appropriate sample size and test duration.

Although these forms of testing are most common in the electronics industry—where accelerated stressing is easily simulated using increased temperature and computer-simulated cycling—they are nonetheless important in other industries. TV advertisements for car makers frequently say that their new models have already been exposed to several million miles of test-track testing. This form of testing is better than no testing, as mentioned above. However, one hundred cars that have been driven 10,000 miles on a test track is not the same as ten cars that have been driven 100,000 miles in city driving conditions. Valid forms of accelerated testing closely simulate the conditions under which the asset will be used; extreme stresses focus on elements most likely to suffer from those stresses.

Accelerated testing of some kinds of systems and components may be difficult. However, all systems contain components for which this testing is commonplace. There are few systems that are not computer- or microprocessor-controlled, with electronic elements stretching across their breadth. It is important to ensure that unproven elements, where failures can produce dramatic effects, are thoroughly tested.

A recent example of pre-production testing involved an enclosure designed to contain a number of different antennas. These antennas were intended to serve the various communications-based systems on modern locomotives. The number and differences in configuration of antennas had grown so large that a common enclosure for the antenna farm was deemed to be necessary. Once a final configuration for the enclosure was found, it would be applied to several thousand new and existing locomotives. Therefore, the design needed to be correct.

The enclosure was to be placed on the top at the front of locomotives. As a result, the exposure to wind, rain, and all sorts of weather would be significant. Although systems would be maintained at intermediate points in time, it was hoped the enclosure would provide a sound and secure environment for relatively lightly constructed antennas for 30 years. During that life, access panels would need to be opened and closed at some intervals; the panels needed to be capable of being safely maintained and resealing after maintenance.

Clearly the challenge in this testing was to identify the extremes of operating conditions and to integrate those conditions in realistic tests. One extreme is the maximum temperature that can be reached when passing through long tunnels (as high as 200° F) and increases with height in the tunnel. It was important to define a regimen of tests that simulated all the extreme conditions and extremes that tended to occur simultaneously.

Several other tools that can be used during this phase of DFR include the following:

ESS—ENVIRONMENTAL STRESS SCREEN

♦ ESS is another form of testing that focuses on environmental stressors like extremes in heat, cold, moisture, humidity, etc. As might be expected, the first step in this form of analysis is to identify what extreme conditions exist and then find a laboratory that can conduct realistic tests.

RDE—ROBUST DESIGN EXPERIMENT

♦ Some products must be designed for misuse as well as extremes of normal use. Tests for robustness in design typically focus on forms of utilization that the product has not been directly designed to endure. For instance, if it is likely that an adult may use a product intended for children, the product might need to be tested to an adult's weight.

RGT—RESULTS OF RELIABILITY GROWTH TESTING

♦ Failures can take several different forms. They can show infant mortality or failure soon after they are placed in service. In this case, if components survive the initial period, they are likely to experience a long and reliable life. They can still experience random failures across the entire life of the asset. Although a few random failures may be acceptable, too large a number is not acceptable. A third kind of failure is wear-out or end-of-life failures. This kind is viewed as the most common and is often accepted as a normal part of life (although it is not).

To properly maintain the asset and to harvest all the inherent reliability, it is important to understand the failure modes and how and when components will fail. RGT is a methodology by which the various failure modes are identified and addressed so reliability performance can improve or grow based on correcting or eliminating the problems found during testing. A good way to think about RGT is that reliability performance is seldom static. If left to the devices of nature, reliability will degrade over time. If appropriate testing and continuous improvement is applied, it is possible to work out the bugs before they begin causing problems that reduce reliability.

BURN-IN

◆ Burn-In is one of the ways in which components can fail. It is the way some manufacturers choose to deal with infant mortality. Burn-in can be used as a screen to eliminate those items likely to experience infant mortality. Burn-in is a form of HASS.

A recent example of this kind of testing involved locomotive traction motors that were manufactured by a shop in foreign country. The manufacturer did not have any direct experience producing this product. Therefore, some form of proof testing was needed before the items could be applied to general use. As with many situations, there was no simple way to create realistic tests that represented the actual working conditions. As a result, a test consisting of applying a specific number of the devices to actual conditions for a specific time was agreed. Until that test was successfully completed and results analyzed, the units from the new manufacturer could not be used.

Apply What You Have Learned

A great deal will be learned when going through the DFR process. The value of accomplishing DFR during the design comes from making changes that will enhance reliability of the finished product. Some changes will involve the manufacturing or construction process. Other changes involve suppliers of components. Still others affect the quality control process being used. Wherever the opportunity, it is important to make the needed changes discovered during the DFR process.

Apply Key Reliability Elements to the Manufacturing Process

Working through all the steps in the DFR process up to this point is likely to provide sellers with a number of insights that must be applied during the manufacturing process to create a reliable product. Here are a few examples of the kinds of findings that might be uncovered:

USE A SPECIFIC SUPPLIER FOR COMPONENTS

If only specific suppliers have been proven to provide components critical to the reliability of the asset, the source of those components should be limited to the proven supplier. It is possible the seller may want to go through the process of qualifying additional suppliers, but until the performance of the components coming from the new supplier is proven, supplies should be limited to the proven supplier. Many sellers have highly structured and disciplined Supplier Qualification processes to ensure these steps are taken every time a supplier is changed.

INSPECT OR TEST THE COMPONENTS RECEIVED FROM A SPECIFIC SUPPLIER

Some suppliers provide the only source for a specific component, but remain a questionable supplier. In this case, it will be necessary to inspect all parts received from that supplier. Even products from trustworthy suppliers should be subjected to inspection on some statistical basis.

FURTHER PROCESS THE PARTS COMING FROM A SPECIFIC SUPPLIER

In some cases, suppliers do not have the capabilities to perform final processing steps for specific components. For instance, a final heat treating step or a final balancing step might be needed. In these situations, it is important to understand the limitations of sub-tier suppliers and take steps to address those limitations.

APPLY HASS OR HASA TO COMPONENTS COMING FROM A SPECIFIC SUPPLIER

In some cases, it might be necessary to place incoming supplies under rigorous testing to screen our weak components before they become a part of the final product. It is important to apply this testing even when critical components are in short supply and supplies will be made even scarcer when tested in a destructive manner.

PERFORM BURN-IN TO THE COMPONENTS COMING FROM A SPECIFIC SUPPLIER

The same comments apply as with the last item.

PERFORM PMI OR MATERIAL TESTING TO THE PARTS COMING FROM A SPECIFIC SUPPLIER

If there has ever been evidence that a specific supplier has substituted incorrect material or had lapses in quality assurance procedures, all materials from that supplier should be verified before use. The cost of performing the quality checks should be added to the costs of components purchased from this supplier.

PERFORM PMI ON ALL COMPONENTS MADE OF A SPECIFIC KIND OF MATERIAL

Scarce or high value materials should be checked to ensure substitutions have not been made.

PERFORM INSPECTIONS AT A SPECIFIC MANUFACTURING POINT

Frequently, "hold-points" in a manufacturing process are needed to ensure that particularly sensitive manufacturing steps are checked before moving onto another step that will obscure earlier steps.

PERFORM TESTING AT A SPECIFIC MANUFACTURING POINT

Similar to the above item, there are occasions when following steps of the manufacturing process will obscure and make testing of earlier steps impossible. Suppose a coating is applied and will be covered with insulation. The coating may need to be inspected or spark-tested before being covered.

PERFORM SPECIFIC INSPECTIONS TO FINISHED PRODUCTS

Before a product is delivered, a completed product should be thoroughly inspected for defects by someone who is accountable only for finding defects.

PERFORM TESTING TO FINISHED PRODUCTS

Before a product is delivered, the product should be tested through the complete range of functionality. If some aspect of functionality is at question, a much broader regimen of tests should be applied.

Initiate and Follow Through with the "Living Program"

After products have been delivered, it is important to institute a "living program" that closely monitors performance in live service. As part of the living program, the performance of the overall asset should be compared to the required macro-reliability (or reliability of the overall system). In addition, failures to individual components should be tracked and compared to the expected performance of those components.

Although the seller's access to information is limited once the asset is delivered and, even more so, after the warranty period has lapsed, it is important to remain in close contact with the asset. Suppose owners are asked what their experiences have been with a specific make and model of an asset. If the sellers respond, "We haven't received any complaints," they are actually saying they have little or no feedback from the owners. This should serve as an indication that the sellers have no effective living program.

An on-going relationship between sellers and the products they sold in the past offers benefits to both the owner and the seller.

♦ The sellers know all the assumptions that were made during design. They understand how each component was expected to work and how each component was expected to fail. If it fails differently than expected, the sellers are best able to shortcut investigative steps that will simply conclude that the device is performing and failing as expected. In other words, if you want different performance, you need a different device.

♦ The last year I spent in the Air Force was in South Korea on a one-year tour of duty. That time was roughly the twentieth anniversary of the U.S. occupation after the Korean War. I recall a discussion where one person commented "The U.S. has been in Korea for twenty years." Another person corrected the first saying, "The U.S. has been in Korea for one year, twenty times." The point he was making was that a series of independent one-year assignments produced far less progress than if experiences were allowed to build on themselves. The same issue is true if sellers do not maintain an on-going relationship and some accountability for their products. They learn anew every time they create a new product. The process is not continuous.

Chapter 9 provides a detailed discussion concerning how data can be collected and used to clearly understand Failure Modes and failure rates, and then link that information to the performance that has been guaranteed by the seller—or at least described in the seller's technical data. Ultimately the burden will be on the owner to prove that an asset or component is not achieving the required performance and that poor performance is not the result of poor operation or maintenance. Most of the data needed to provide such proof must be collected in a well-structured and highly disciplined data collection system. An example of such a system is described in Chapter 9.

Conclusion

At the conclusion of the steps described above, it is expected that the sellers will have produced a reliable and profitable product. It is also hoped that the owners will have received an asset that fulfils all of their needs and provides a useful tool in producing the products or services that are the focus of their business. Although it is hoped that both objectives will be achieved, there is no guarantee. Because the sellers have more influence than the owners in the way these steps are accomplished and the manner in which success is measured, there is a far greater likelihood that the sellers' objectives will be achieved than the owners' objectives. The sellers' objective is to create and market a profitable product; that objective is the most likely to be achieved when the DFR process described in this chapter is applied.

The next chapter describes a process designed to deliver on the owners' objectives. Unlike the sellers' objectives, the owners' goal is to procure an asset that will become a long-term part of their process for producing products or providing services to the owners' customers. Owners are looking for different information and different results from the DFR process. Although many elements of a DFR process for owners can be very similar to the DFR process used to achieve the sellers' objectives, there are differences. If owners do not address those differences in their specifications and do not obtain assurances their requirements will be met before agreeing to purchase the asset, it is likely they will be dissatisfied with the results.

Chapter 3

DFR for the Owner's Business Model

Efficiency is doing things right; effectiveness is doing the right things.

Peter F. Drucker

As the buyer and long-term owner of an asset, my needs and objectives are somewhat different than those of the seller. Although I certainly believe that sellers should do everything in their power to produce a reliable product, there are some aspects of the manner in which they develop a product to suit their business model that does not suit mine. As the buyer, I need to be aware of those differences and make certain the sellers will accommodate my needs **before I agree to purchase their product.**

Macro-Reliability and Micro-Reliability

Although we will provide a detailed discussion of DFR for the owner's business model, let's begin with a high-level description of a way that the owner's needs are different from the seller's.

Most owners are concerned only with the *macro-reliability*—the reliability of the overall asset. Indeed if the macro-reliability of the asset is good enough, the owner may need to be concerned with little else. I recall a situation in which a set of high voltage rectifiers were so reliable that, as owners, we did no maintenance and replaced no parts for a period as long as ten years. Our only concern was that when a breakdown ultimately came along, we would have no experience servicing the rectifiers and the needed parts would be obsolete. Unfortunately, there are few situations in which the absence of failures or the small amount of maintenance allows the owner to get out of practice.

More often, the macro-reliability of an overall system is far from perfect. As a result, there are times when it is necessary to perform diagnosis and troubleshooting to identify failed components. For purposes of this discussion, we will refer to the performance of individual components as the *micro-reliability*. The micro-reliability of a system is the performance of the individual components that ultimately adds up to produce the reliability of the overall system (the macro-reliability). When the overall system experiences poor reliability, the analysis will ultimately identify one or more of the components that failed.

As an owner, I need to understand what should reasonably be expected from the standpoint of both macro-reliability and micro-reliability. When my overall system is not performing up to expectations, I need to quickly identify which individual components are not performing as expected. Rather than spending needless time trying to sort through a number of components that have failed, some of which are meeting expectations and some of which are not, I need to quickly attack the components that are not delivering the expected performance.

Although one part may have a higher failure rate than another, it may still be providing all the reliability that was reasonably expected of it. The poor macro-reliability of the entire asset may be the result of a part that is not the worst performing part, but is still not performing up to snuff. The owner may have little leverage to force the improvement of the part with the poorer performance. However, the owner significant leverage to demand action that improves the performance of the part with the better performance, but that is not performing up to expectations. Owners who do not demand and receive micro-reliability data will have little ability to address issues at other than the macro-level.

Sharing Information

Generally speaking, OEMs are unwilling to share the kind of information described above. Although they will share the macro-reliability numbers, they are unwilling to share micro-reliability data. There are several reasons for this reticence. First, the level of detail needed to describe the component reliability is also sufficient information for the owner to find a different source for the components. OEMs are

concerned that providing customers with this information will leave them high and dry as a source for future parts sales.

Second, the seller's analysis leading to the macro-reliability is not exact. Some components may underperform; others may over-perform. Occasionally the over-performance of one component may more than offset the underperformance of another. In this case, the macro reliability of the overall system may be as advertised. If owners understand the expected performance of all components, they may demand that all components perform as advertised. In this case, the sellers will be held accountable for poor performing components even when the overall system performs as advertised.

Finally, this kind of detail may uncover a number of the sellers' "dirty little secrets." One secret is that the sellers really don't know as much about their product as they put on. Another secret is that the sellers' control of second-tier supplied components may not be as advertised. In fact, the supplier may be shifting and changing component suppliers frequently. When changes occur, the actual component may not have the same reliability as the component used in the determination of the overall asset reliability. If owners have facts, it may raise their concerns and increase their leverage.

As a result, the analysis and information needed for owners to maintain the inherent reliability of their assets is somewhat different than that needed for the sellers to get the product out the door and delivered to the owner. The only way the sellers will invest the effort required to develop and provide that information is when the owners have clearly stated that requirement in the purchase specifications. The only way that the sellers will agree to provide the owners with all the data developed during the DFR process is when the owners have made it a mandatory part of the purchase agreement.

The issue all gets down to that single difference. The owners need the information required to ensure the inherent reliability of an asset for the entire life of the asset. The sellers must do what is necessary to keep selling products and to make a profit, including the impact of the cost of warranty and post-warranty liabilities.

The simplest ways to describe the owner's requirements for information products coming from the DFR process is to separate the information into the elements associated with Reliability, Availability, and Maintainability. It is important to emphasize again that the asset life described by the owner's business model is much longer than the asset life for products being sold in the seller's business model. The owner's asset life may be 15 years, 30 years, or longer. By comparison, the seller's economic analysis may be limited to the period covered by the warranty. In each of the characteristics described below, the objective will be to quantify the value of losses resulting from underperformance over the entire life span being analyzed. Sellers analyze losses due to underperformance during the short period for which they are accountable. Owners understand that they will have to deal with shortcomings of the asset forever. However, from a practical standpoint, they limit the analysis to the period over which the asset value is amortized.

In addition to the detailed descriptions provided below, an example of a purchase order can be found in Appendix 2 at the end of the book.

Reliability

Owners must be provided with data that clearly shows the asset they are purchasing will provide the required level of reliability over the entire specified life cycle. In turn, the sellers must show that the approach being used has been successful in forecasting the actual reliability performance in past situations. If the method being used can not be shown to be accurate, it will be of little value to the owners.

One method of forecasting the overall asset reliability of a product during the development process is the Reliability Block Diagram (RBD) method.

Reliability Block Diagram (RBD) Method

For simplicity, we will confine our discussion to the simplified manual form of RBD analysis. Although the manual method of applying RBD is useful for analysis of simple systems, the complexity of systems and the attendant analysis quickly becomes too great to handle manually. As a result, a number of software makers have produced programs that support RBD analysis. Chapter 4 will focus exclusively on reliability analysis.

The reliability of any single component can be represented by a block like the one in Figure 3.1.

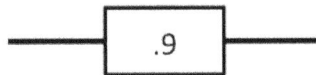

Figure 3.1 A single component in RBD

In Figure 3.1, .9 or 90% represents the reliability of the component. A component with a 90% reliability is expected to have a 90% likelihood of surviving one year without a failure, or a 10% likelihood of failure.

There are a variety of sources of information concerning the reliability of different kinds of components. The most accurate is probably the owners' direct experience. If the owners are currently using a component or piece of equipment that is the same as the one being proposed for the new facility, then the new component should perform with the same reliability as the current one. That assumes the owners will operate and maintain the new one in the same manner as the current one and it will act in the same manner.

A second source of information is the supplier of the component. OEM's data is most believable when the OEM is providing a warranty that the component will perform to that level. In addition, there are a number of books and articles available that provide typical performance for generic components. These last sources do not consider specific operating or maintenance impacts.

Even the most complex of systems can be modeled using a few simple sub-system configurations. Figure 3.2 shows components arranged in series.

Figure 3.2 Components arranged in series.

In cases of series configuration, the combined reliability of the two components operating in series is simply the product of their individual reliabilities. In Figure 3.2, the combined reliability is .72 or 72% (.9 x .8 = .72). The system consisting of these two components would have a 28% likelihood of failure in any one-year period.

Figure 3.3 shows how the components can be represented in the case of a parallel configuration.

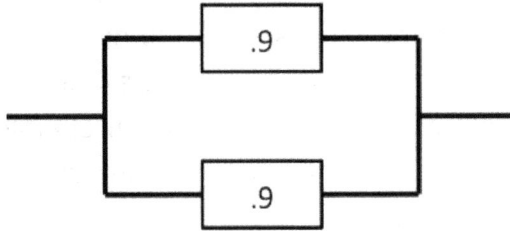

Figure 3.3 A parallel configuration.

In this case, the combined reliability of the subsystem would be enhanced by redundancy or "sparing." The following equation is used to determine the combined reliability:

R = A + B – (A x B)

= .9 + .9 – (.9 x .9) = 1.8 – .81 = .99 or 99%

In other words when a 90% reliability device is spared with a similar device, the combined reliability is 99%. The likelihood of failure is 1%.

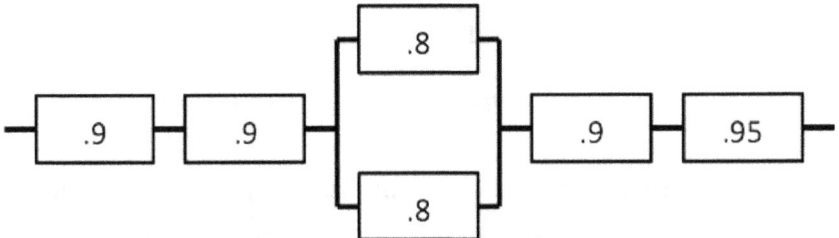

Figure 3.4 Modeling a complex configuration.

As Figure 3.4 shows, these simple subsystem configurations can be combined to model almost any complex configuration.

The math is left to the reader. Despite the fact that all unspared components have what seems to be a good reliability, the overall system reliability is only .665 or 66.5%. Thus, the system has a 33.5% likelihood of failing in any given year. If statistics play out, there will be one failure every three years.

In addition to the features described in the simple models above, it is possible to include the impact of spare parts strategies, intermediate product storage, and various other characteristics using computer simulation software.

Reliability Block Diagram modeling is a complete subject in and of itself. Although it is not my intent to provide a comprehensive treatment of RBD, a few useful details follow:

◆ The objective of preparing an RBD model is to develop a general understanding of the reliability of an asset before design and construction are complete. By that time, owners are "stuck" with what they have. The model must adequately describe the system being studied. The computer model cannot completely capture all the elements. However, it is useful for comparing the elements that have been captured.

For example, when I was a boy, my father owned a number of old trucks. He nurtured and pampered them, and never asked them to perform tasks beyond their limits. As a result, he received excellent service from the old trucks. On several occasions, these trucks experienced major failures (blown engines) shortly after he traded them. This kind of care cannot be included in a model because the model doesn't know who will be chosen to operate and maintain the asset.

♦ RBD software varies in its capabilities. It is important to select software that has the features you need. For instance, one version of RBD software has the ability to simulate the filling and emptying of intermediate and run-down tankage. If you need to understand the capacity needed to allow closely-coupled assets to continue to operate during short interruptions, this feature will be important for your RBD software to contain.

♦ The best information for failure rates comes from your own records (because it reflects how you operate and maintain your assets). However, a variety of useful sources of reliability data can be used to "fill the boxes." For instance, the IEEE Gold Book contains a great amount of data for electrical and electronic components.

♦ As mentioned earlier, the ability to identify very infrequent failures depends on performing enough simulations for the model to produce the low likelihood but high impact events. Almost all forms of RBD software allow users to select the expected life and the number of life simulations that will be included in the results. Users must be patient while the software grinds through enough simulations to produce realistic results. They should begin with a moderate number of simulations, then increase the number and re-run the program to see if there have been meaningful changes. If so, they should increase the number of simulations until the results no longer change by a meaningful amount.

A reliability allocation method is another simple approach frequently used to identify the overall reliability of an asset. This approach tends to reflect the structure of organizations where it is used. The overall reliability is assumed to be additive. Each part of an organization that is assigned the responsibility for specific systems or sub-systems of an asset are provided a reliability allocation or a portion of the assumed overall unreliability that their systems can produce. Suppose the allowed failure rate amounts for one failure per year and an asset is made up of three major systems (engine, control and drive). Then the allocated failure rate might be as follows:

◆ Engine — .5 failures per year
◆ Control— .25 failures per year
◆ Drive — .25 failures per year

As long as each of the departments of the organization stays within their allocation, it is believed that the overall asset will also perform as required. The weakness of this approach is that it does not account for the statistical nature of failures. Although it may produce an acceptable (or even more than acceptable) level of performance in one period, it may produce marginal performance in another.

For purposes of this discussion, there are a variety of ways to forecast the reliability of future products. The objectives of applying these techniques are twofold:

1. The techniques are intended to show that the product being developed will meet or exceed the owner's specified requirements.

2. They can be used to highlight areas of weakness so changes can be made in an iterative manner and the performance can be made to meet requirements.

No matter how sophisticated the analysis, if it cannot be shown to generate an accurate forecast, it is of little use.

In the two examples above that are used to forecast reliability for the asset (earlier called the macro-reliability), it can be seen that overall asset reliability is related to the sum of its parts. This is an important albeit subtle point. It is important because it highlights the owners' need to understand the reliability of individual components (micro-reliability of the components) as well as the macro-reliability. It is subtle because all too often we overlook the negative impact a fairly innocuous component can have.

While I am in the process of writing this book, the world is hearing more and more about a reliability problem with several models of Toyota vehicles. As the story unfolds, there are a series of accounts about the part the floor mats have played in this problem. The problem is that the vehicles unexpectedly accelerate in an uncontrollable manner. The first round of stories said that the floor mats were the cause. As the newspaper accounts progressed, they said the floor mats played a partial role.

This story has a long way to go before it is over, and I am not suggesting this discussion is addressing the final story. Still, it provides a good example of the

issue described above. If owners received an estimate of the macro-reliability and micro-reliability of the subject automobiles, would that analysis have named the floor mat as a possible contributor to an acceleration problem?

If so, how would have owners responded? They would remove the floor mats at least until they were certain they did not introduce an unnecessary risk. It is only in a world where owners do not recognize that a significant risk exists that they allow this relatively meaningless issue to continue to play an unmanaged role in the safety of themselves and their families.

The removable floor mats are probably an extreme example of the issue. However, when owners purchase a capital-intensive, long-life asset, they should be able to expect that the sellers will provide an accurate forecast of the overall or macro-reliability as well as the micro-reliability of all components used in creating that forecast.

The information provided by the sellers should include the following details:

♦ Contributing elements versus non-contributing elements
  • Components included in the calculations are viewed as having failure modes that will contribute to the overall life of the asset
  • Components not included are viewed as having no failure modes that will contribute negatively to the overall asset reliability
♦ Failure Frequency and usable life (Weibull life)
  • Include the expected reliability of each component
  • Also include the usable life over which the component

will deliver the expected reliability. If the component must be replaced at some time to maintain the overall asset reliability, the owner must be made aware of this and the cost of such replacements must be included in the TCO.

♦ Failure Mode or Modes included and characteristic of each Failure Mode

- • If more than one Failure Mode exists, they should all be included
- • If more than one form of predictive or preventive main tenance is needed to ensure reliability or extend the usable life, the owner needs to understand all of them
- • Be sure to include all forms of risk mitigation tasks or sys tems (e.g., operator inspections, actions imbedded in procedures)

The ability to make informed choices between higher first costs and higher costs for the entire life of an asset is only possible when owners have all the information. Lacking this information, it has to be assumed that the seller is making decisions for the owners, which further assumes that the seller knows better than the owners what choices are best for them.

When a specific component has been excluded and is therefore viewed as having no failure modes that contribute to poor asset reliability, any failures are unacceptable. (This point is important to the Toyota floor mat example. Before they became a possible explanation, were floor mats ever viewed as a possible cause for fatalities? Had they been, would Toyota ever have provided them or would they simply have avoided the responsibility by allowing interested owners to provide their own optional floor mats?)

The ability to quickly separate unusual failure rates and Failure Modes from those that should be expected is possible only when owners are provided information describing those expectations and when they track asset and component performance in a manner consistent with the information that has been provided to them.

Availability

Next, owners should ensure that Availability is addressed as part of the design. Simply put, Availability is a measure of the portion of time an asset will be able to perform its intended function. If the function of the asset is to make a specific product at a specific rate, at a specific quality, and with a specific efficiency, then any interruptions in performing that function or any aspect of the function is a loss in availability.

In addition to considering all the characteristics that are important to functionality (production rate, quality and efficiency), it is important to consider both forms of interruption: planned and unplanned.

Planned interruptions are those that are expected and typically the result of issues beyond the control of the designer. For instance, it is possible that heat exchangers or some other element tend to foul or become less efficient the longer they are in service. Assuming there is no other efficient and effective way to address issues like normal fouling, the asset will need to be idled for cleaning on some regular cycle. Routine cleaning and other similar activities come under the heading of a planned interruption and should be understood by the seller and the owner during the design. Regular outages for overhaul and renewal make up another form of planned interruption. There are a variety of names for this kind of event: shutdown, turnaround, overhaul, and outage. Whatever the event is called, the characteristics are as follows:

♦ It typically happens every several years as limited by one or more components that have reached the end of their useful life.

♦ It typically takes much longer than other forms of mainte-nance as limited by the longest critical path duration of one or more components that must be maintained during the event.

♦ It typically renews all elements of the asset that cannot be readily maintained while the asset is operating.

As might be expected, the losses associated with periods of planned unavailability are some of the most significant during the life of an asset. These losses are critical to the owners' prof-itability and play a key role in their business model. It is absolutely critical that sellers understand this characteristic and clearly communicate it to the owners.

Suppose some owners purchases an asset that produces a specific product. The owners are the only company that makes that product and that several other companies depend on the owner for a supply of the product. Those companies make commitments to their customers for their products, so ultimate-ly an entire business chain depends on the original owners' ability to produce products. The owner discovers that their asset requires an overhaul much sooner than expected. Furthermore, the overhaul will take much longer than expected. The owners had plans and supplemental capacity to overrun demands in periods leading up to outages so they could build inventory in anticipation of the outage. The owners also had built storage facilities to keep additional supplies to be dispensed during the outage.

But now the owners learn the outage is coming sooner than expected; there is no time to build up inventories. Even if there was time to build up inventories, the storage facilities would be too small to cover the entire outage period. As a result, the return on investment will not occur for the additional facilities the owner built to provide added capacity and storage. In addition to investments with no returns and loss income due to early and extended outages, the owners may be subject to legal action and to loss of customer loyalty because of their inability to supply products as committed. Early, unexpected, and extended outages create situations that are bad for business in a number of ways.

To a lesser extent, instances that allow an asset to continue to operate—albeit at lower efficiency or a reduced rate— should also be known, described, and communicated by the sellers to the owners. In these situations, the owners need to accommodate the issues in the design capacity of the asset, the storage available, the expense budget (e.g., cost of additional energy during periods of inefficiency), and the profit forecasts. Not knowing about inherent physical limitations of an asset places owners in jeopardy of creating a situation that does not fit their business model or the story they are telling their investors.

The other portion of unavailability is the unplanned portion or the part that is the result of poor reliability. Although this part of unavailability cannot be quantified to the same level as planned unavailability, it can be quantified to the degree that the level of understanding will support the owners' business model and their business. Said another way, much of the DFR analysis is aimed at identifying failure modes and their effects. It also identifies the tasks and the duration of the tasks needed to recover from those failures. When that information is provided to the owners, they can determine if the frequency or duration of those interruptions or the necessary response is consistent with the way their business operates.

Too frequent interruptions (low reliability) will consume all the attention of plant leaders. Poor maintainability will result in long outages or unstable restarts. This condition may make it impossible to assure customers of the product they require. Once the owners are duly warned of anticipated reliability performance, they may say that these conditions are inconsistent with the manner in which their company does business. Therefore, the defects causing this behavior have to be eliminated before the design can proceed.

Two factors are needed to quantify unplanned unavailability that must be determined. The first is provided when the reliability model of the asset is complete and simulations have been run. These results will tell the seller and the owner how frequently an unplanned event resulting in the loss of availability will occur. If a manual model has been used to determine the expected reliability, the product of that analysis will be a statistical likelihood of failure in any given year. In other words, if the reliability of the asset is 99%, there is a 1% likelihood of failure in any year. Rather than zero or one, the expected unplanned unavailability should be .01 times the interval of the likely event. If a computerized model has been used, the results will identify the specific components that will fail in any given period.

This brings us to the second factor of estimating the effects of unplanned unavailability—the expected duration of the unplanned event. This information will be made available from another part of the analysis. When performing the work necessary to ensure Maintainability, it will be necessary to perform a streamlined RCM analysis. The streamlined RCM will identify the typical tasks that need to be accomplished over the life of the asset. Once all the expected maintenance tasks have been identified, sellers will need to analyze each task. The object of the analysis will be to answer the following questions:

1. Can the task be accomplished in a ratable period of time?
2. Will the ratable task restore the full inherent reliability of the asset?

We will discuss this activity more in the next paragraphs. Having performed a thorough analysis of maintainability will deliver all the information needed to estimate the typical outage duration. In performing this analysis, it is important to include the duration of all peripheral activities that may be required when an unplanned reliability-related event occurs. The surrounding activities—like decontaminating or isolating the asset—may actually require more time than the actual repair. Although the likelihood of an event might be small, the duration of the event might be so large as to make the annual allowance for unplanned unavailability fairly significant.

If a manual approach has been used to model reliability, the duration of the event used to determine unplanned loss of availability will be the duration of a "typical event." Therefore, it is important to include all the peripheral activities. In many cases, the duration of a typical event has more to do with peripheral issues (taking a locomotive to a shop or shutting down and decontaminating a process plant) than it does with the actual repair. However, if a computerized modeling system has been used to simulate failures, it will be possible to apply the specific duration of the repair associated with the failures forecasted by the model.

Again, it would be unwise for owners not to include this element in their business plans. Furthermore, it would be unprofessional for the sellers to not tell the owners about these statistical, but still expected interruptions.

Maintainability

The subject of maintainability was introduced along with availability in some detail. In many ways, the two characteristics are inseparable. If an asset is unmaintainable, it is likely to have poor availability. The maintainability of an asset will determine how long it will be out of service when it fails and if it is likely to fail again shortly after being maintained. Maintainability is defined as a measure of the capacity to restore the inherent reliability of a system in a ratable period of time. A useful measure that can be used to quantify this characteristic statistically for a number of devices or components is the Mean Time Between Failure (MTBF). To quantify the maintainability of an asset, begin with knowing the tasks that need to be assessed for those properties. Start with some process for identifying both the components that will be likely to fail and require repairs, and the predictive and preventive maintenance needed to minimize the number of such repair tasks. The most commonly used tool to provide that information is Reliability Centered Maintenance (RCM). In order to perform RCM in the most effective and efficient manner possible during the design process for an asset, it makes best sense to select a streamlined RCM process.

RCM is a Failure Modes and Effects Analysis (FMEA). It has an objective of delivering all the inherent reliability of an asset by identifying the tasks needed to mitigate or eliminate risk of failure resulting from known failure modes. When conducted as part of a DFR activity, an RCM analysis will serve the second purpose of identifying required maintenance activities. RCM will identify the reliability of each component that should be expected when properly maintained. This provides a second estimate of component reliability. This estimate can be compared to the value for the reliability of each component that was used during the RBD analysis to calculate the overall asset reliability. If the reliability produced based on application of a reasonable maintenance program is inconsistent with the values used in conducting RBD, the differences should be reconciled. If the component reliability (micro-reliability) is too low, the desired system reliability (macro-reliability) will never be achieved. If the component reliability is higher than necessary to provide the required system reliability, the amount of maintenance being recommended may be greater than necessary.

There are a number of different forms that RCM and streamlined RCM can take.The difference between RCM and streamlined RCM has consumed a great deal of attention over the last few decades.In the author's experience, all forms of RCM are just a starting point.Ultimately the tasks being done must be optimized in the years following RCM using a "Living Program" that continually evaluates both the cost and the effectiveness of the regimen of work.As a result, I believe it is best to begin as quickly as possible and do so with the smallest initial investment.Those characteristics will result from streamlined RCM. For the purposes being described in this text, the RCM analysis is being performed

Once the RCM analysis is complete, the resulting component reliability is reconciled, and the maintenance tasks are identified, the engineers involved in developing the new product should walk-through or simulate each task in a manner that will determine the task efficiency and effectiveness.

Task efficiency is another way of describing the tasks ratability. Can the task be done in a ratable period of time? Or are elements of the task unsure, involving steps that may require multiple tries? Are there tasks that require mechanics to stand on their heads or squeeze into unreasonably tight areas to complete the repair? Any of these circumstances may result in a task for which it is impossible to set an accurate duration.

Task effectiveness is another way of describing the ability of the task to restore the inherent reliability of the asset. If the results at the end of the task are unsure, then task effectiveness is in question. Consider the case where multiple electronic components are simply changed one-by-one until the functionality is restored. In this situation, there is no way of individually testing the electronic components to determine if they are good or bad. The defect may be identified or it may not. Removal and replacement of several components may simply "jiggle" things enough to restore poor contacts. These same marginal contacts may quickly fail again after the asset is placed in service.

Think about the last time you called for computer support. The help desk probably asked you to recycle your computer by turning it off, then back on. It is reasonable that this got things working again for you. But did it repair the problem? Although the fix worked temporarily, it did nothing to restore the inherent reliability of the system.

Format

Although there are a number of ways to deliver the information, the most important issue is that the information is readily usable and not confusing.

For instance, there are a variety of ways to report the reliability of electronic components. The most significant difference is the time span covered by the reported likelihood of failure. A device is much more likely to fail in any given year than it is in a single day. Also, the likelihood of failure for some devices tends to increase at the end of their usable life. This characteristic would create a requirement that the component be replaced on or before the time that it can no longer support the required reliability for the overall asset. The information needs to be provided in a way that owners can clearly understand the requirement. The important information cannot be hidden in rows and columns of numbers that have to be interpreted to understand their implications.

Typically, the owner will be able to measure failure rate in large increments of time. A large increment of time may be a month, a year, or some other period used to evaluate quantity passing through the inventory system. The data should be provided by the seller in a manner that can be easily sorted for intelligence; it should not require a great deal of manual analysis to use. The data provided should not be open to differences in interpretation. Therefore, when a difference between estimated and actual is identified, the solution does not have to begin with a long discourse over the meaning of the data.

The products coming from Weibull analysis provide a good basis for the kind of information that will be useful. The information should be provided in an easily sorted database. Then, when failures begin to occur, it will be easy to find the data for the failed component. It will also be easy to interpret the performance that should be expected. This point will be discussed at greater length in Chapter 4.

Timing

It is expected that sellers may have a million excuses why the results of the reliability analysis and the specific data requested by the owners cannot be delivered until some time after the asset has been delivered. Still, that is not good enough. Suppose the data is not available until after the owners have taken possession of the asset. That fact alone would serve as a clear sign that the sellers have not applied the results of the analysis in the design of the asset. After the owners have taken delivery of the asset, it will be too late to identify shortcomings and demand changes.

It is important that the owners' reliability representatives be ever present at the designer's office while the asset is being designed. They can then observe DFR in progress and review the results of the analysis in real time. The data provided by the sellers to the owners at the time of delivery should not be new to the owners. It should only be data the owners' reliability representatives have seen and approved earlier, but packaged in a neater, more usable form.

The absence of on-site representatives will lead to choices being made to meet the sellers' needs through the design process. In order to ensure that the needs of the owners' business model are being addressed during the design, a knowledgeable reliability engineer must be on-site at the designer's office from the start of design to the delivery of the asset.

Application of the System Engineering V-Model

The last chapter introduced the V-model for system design. The V-model (Figure 3.5) provides a helpful tool for understanding:

- The general activities that should be included in the design and development process.
- The typical order of those activities.
- The nominal timing the key activities can be expected to be accomplished.

Although it is unlikely that any established sellers will total-

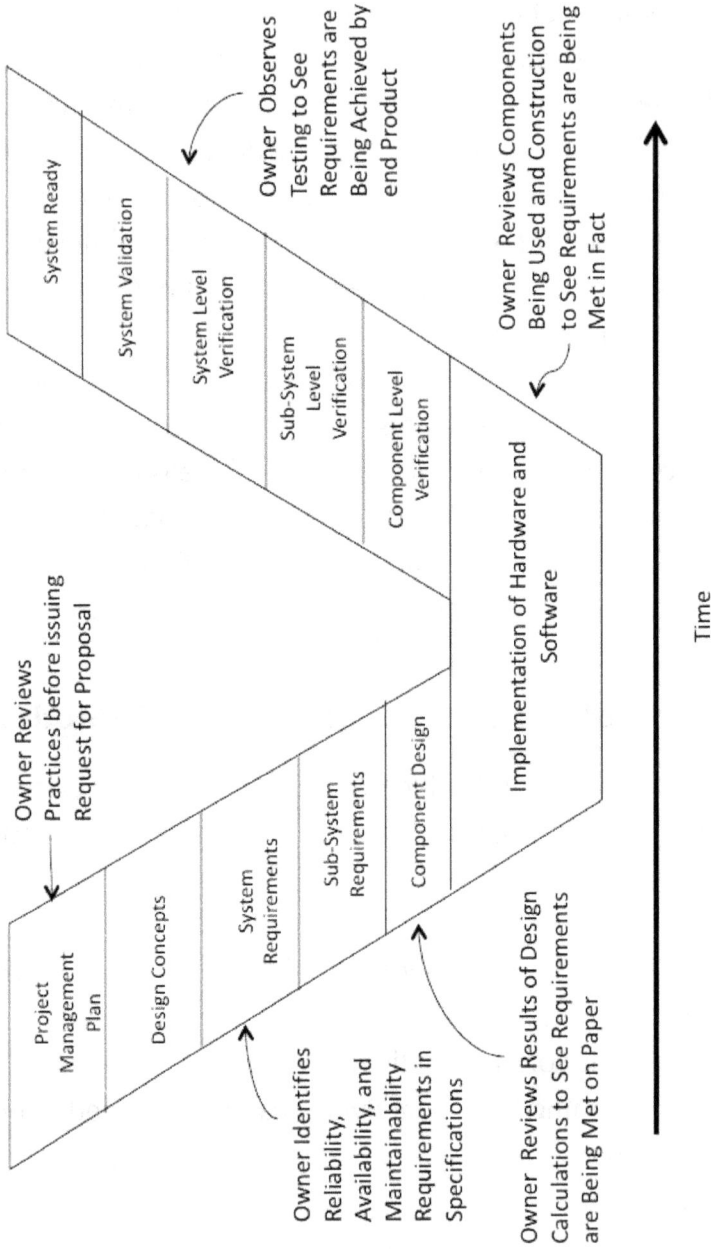

Figure 3.5 The System Engineering V-Model

ly change their design process to meet the needs of their customers, it is likely that the customers can have a positive impact if they understand the process and make certain they are in the right place at the right time during design.

In some cases, the owners may feel that the performance of a product is so important that they will assign embedded personnel for the entire period the design and development is being done. Others may not feel the same drive for continuous involvement. Still others may wish to participate, but do not have the resources needed to provide continuous contact.

In any case, it is important to understand the V-model and determine from it the specific actions the owner can take to have an impact on the final product. The following version of the V-model has been modified to show some of the specific activities the owner's representatives can perform to impact the final product. If the owners can work with the sellers to identify the timing of each element of the V-model, they can be certain to have their representatives there to participate in a timely manner.

For the overall DFR process to be effective, it is important that the individuals assigned to perform the customer and owner activities clearly understand the desired objectives of each step of the process and what actions must be taken to achieve those objectives.

Take, for instance, a typical contract inspector or, for that matter, even a full-time employee being used as an assigned inspector. How many of the individuals assigned to those roles actually understand your company's business model—and how assets should be designed and constructed to fulfill that business model? Realistically, the answer is very few, if any.

Most of the inspectors I have worked with during my career tend to take a very pragmatic view of the world. That view is based on the experiences they have had over their career. Their pragmatism may lead them to believe that an asset with a life span of thirty years and a high level of reliability over that entire period might be unrealistic. They may allow those pragmatic beliefs to prevent them from holding a supplier's feet to fire when providing a product. In the majority of situations where DFR is being installed and used for the first time, the owners' objective is to make a significant change in the reliability of the products they are purchasing.

In this case, the on-site inspectors' viewpoint must be based more on the future requirements than the past problems. Their past experiences should not be allowed to impact future expectations.

Although I used contract inspectors and construction inspectors as examples of individuals who may not share a vision of an asset with a dramatically improved level of relia- bility, this weakness is probably more prevalent among busi- ness and engineering professionals who have little or no back- ground with reliability engineering. Individuals who do not understand that reliability, availability, and maintainability are choices rather than consequences are likely to feel resigned to accepting whatever the seller offers. When the seller tells them that it is not good economics to build highly reliable assets, they believe it. When sellers hearken back to "the good old days"—when assets reliably failed once every few months "and you knew exactly what to expect"—these individuals have lit- tle knowledge about how much has changed and performance from their "good old days" can no longer be accepted.

It is critical that the individuals assigned by the owners understand what is required at each level of the design and development process and be unwilling to accept short cuts, marginal components, or performance that is less than specified. It is likely that the level of performance portrayed by the sales representatives before the sale will be viewed as an exaggeration by the engineers and designers when the actual product is taking shape. In this situation, individuals assigned responsibility for oversight by the owner needs to understand:

♦ What was specified
♦ What was sold
♦ What was purchased
♦ What will be accepted and what will not

If the final product is not likely to fulfill the requirements, the owner's representative needs to raise a flag?and the sooner, the better.

Conclusion

Although the DFR process needed to meet the owners' objectives is different from the DFR process for the sellers' objectives, they are not mutually exclusive. By specifying what they need to meet their objectives, owners are not relieving sellers of their responsibility to perform all the steps needed to meet the sellers' objectives and to create a reliable and functional asset. For instance, owners may not need to understand or be involved in the sellers' process for ensuring the capabilities and quality process for second- and third-tier sources. The fact the owners do not demand information covering these activities does not relieve the sellers of their responsibilities concerning those areas.

Owners need to be provided with a proof statement ensuring that the asset will provide:

♦ The required level of reliability
♦ The required availability
♦ At the expected LCC
♦ For the required life

Each of the characteristics that are described in the proof statement provided by the sellers to the owners are items that are critical to the owner's business model. They also require a form of analysis that is beyond the analysis that the sellers performs for their own needs. As a result, I have attempted to limit the additional activities needed to satisfy the owners' DFR requirements.

Chapter 4

Reliability Analysis

It is better to understand a little than to misunderstand a lot.

Anatole France

In this chapter, we will focus on reliability analysis as it specifically applies to serving the owner's interests during DFR. Current literature tends to focus on DFR methodology that can be applied in a manner that will support either the owner's needs or the seller's needs. The problem for the owners is that the methodology will not be applied in a manner that fulfills their needs unless specifically demanded and enforced by the owners. To be successful, the owners need to know what to specify, how to specify it, and how to enforce his demands.

Applying RBD

Let's begin with a simple example. Chapter 3 provided a detailed description of the Reliability Block Diagram technique. Think for a moment about the numbers shown in each of the blocks (or applied to modeling software). Each number represents the reliability of a specific component. Depending on the form of RBD being used, the number in the block may be a constant (as in a manual calculation) or it may change and be renewed (as with computerized models and simulations).

For the reliability to be viewed as changing over the life of the asset, it will be necessary for the people performing the RBD to make some assumptions. First, the component is always maintained at exactly the right time. Therefore, the reliability is the same as new or it can be assumed that the reliability may reflect more realistic expectations for replacement. In a real-life situation, inspection accuracy may be less than perfect. Some maintenance may not be as timely or effective as hoped. As a result, if the sellers includes an assumption of perfect maintenance in the RBD, the resulting reliability may be unrealistically high. Unless the owners are willing to conduct frequent and unrealistically expensive maintenance, the actual reliability will be less than forecast by the sellers.

Example: Pressure Sensor

Consider this generic but real-life example to further analyze this issue. A component commonly used in complex systems is a pressure sensor. Modern pressure sensors consist of a housing containing a metal or ceramic diaphragm. An electronic element is mounted on the diaphragm; this element produces a changing signal as the diaphragm is flexed. The diaphragm flexes with changes in the pressure of the fluid being measured. As a result of this arrangement, it is possible to determine the pressure of the fluid in a system.

Several other features of the physical system are important to understand. The housing that contains the diaphragm typically takes the shape of an annulus that leads from the source of the fluid to the diaphragm. Real-world needs for mounting this device typically result in that configuration along with a threaded section that can be mounted on the housing of the device where measurements are taken. Another subtle characteristic is that the various parts of the pressure sensor are frequently made from different materials that best serve the various functions. Specifically, the housing and the sensor may be made from different metals.

The physical arrangement and construction of a typical sensor can provide a host for several different failure mechanisms and, therefore, lead to several different failure modes:

1. Fatigue of the diaphragm as a result of flexing can lead to cracking.
2. Plugging of the annulus with debris can make the sensor inoperable.
3. Concentration cell corrosion due to plugging or corrosion resulting from metals with differing electrical potential can either add to plugging or can weaken the materials in the pressure sensor.

When performing an RBD analysis, the reliability of the pressure sensors that are part of complex systems may play a significant role in several ways. Partial failures may lead to incorrect and misleading signals that may cause the system to operate improperly. Complete failure of the pressure sensor might cause in a complete loss of signal, resulting in shutdown or loss of an important protective function.

Depending on the importance of the function provided by the pressure sensor, the designer working for the seller may need to assume that the device is 100% reliable. Implicit in that assumption may be further assumptions:

♦ The device is always changed out before the end of its fatigue life.
♦ The device is cleaned at regular intervals to ensure the annulus is not plugged.
♦ The fluid being measured never contains materials that are corrosive to materials within the pressure sensor.

Although these assumptions may be completely correct, it would be safer and more realistic to assume they are not. All too frequently the individuals responsible for the designs operate in a pristine world that provides them with little exposure to the realities of life. Their lack of understanding leads them either to a reliability that is far less than advertised or to a maintenance regimen the owners cannot afford to support.

At the beginning of this chapter, I provided the quote: It is better to understand a little than misunderstand a lot. That saying has particular importance to the field of reliability. All too frequently, components or sub-systems are designed in a manner that go well beyond the direct knowledge of the designer. For instance, an electronics engineer may design an element that should have had a metallurgist involved, but did not. When that occurs, the resulting product is likely to be unreliable—unless the designer is particularly lucky. I prefer not to depend on luck to produce the needed reliability.

Another useful quote comes from Albert Einstein: *If you can't explain it simply, you don't understand it well enough.* This saying provides an important insight into the value of reliability analysis during design. The need to create a model of a system used in performing the reliability analysis forces the designer to describe the system, its functions and interactions, and the expected reliability of components in a very simple way. Systems that cannot be described in a simple manner frequently depend on some pretty sketchy theory for operation. That is bad practice.

DEFINING DFR

With all the preceding as an intro-duction, exactly what is DFR—or relia-bility analysis that is conducted in con-junction with the product design process? Basically DFR is a system that describes how the overall system is likely to fail if it is operated and maintained as expected and is exposed only to the forms of deterioration assumed in the analysis.

Unfortunately, the real-world requirements for the system are frequently different than those assumed by the DFR if the analysis is accomplished in a naïve or one-dimensional manner. In order to avoid the overly simplistic approach, DFR must assume that:

- The device is not expected to function in the required manner just once so the seller can push it out the door. It functions in the same effective and efficient manner for its entire life.
- The device is not expected to function only under optimum conditions; it functions in all possible environments.
- The device doesn't just function in the manner that the designer pictures for it to function; it functions in all the ways that nature will cause it to function and that it will fulfill the owner's requirements in all those situations.

These issues tend to separate DFRs conducted from the seller's perspective from those conducted from the owner's perspective. The terms "robust" and "real-world" will likely mean different things to owners and sellers. Sellers may build a product for what they interpret as the demands of the general population. Owners purchase a product to meet the specific demands of their business.

Earlier chapters described examples of how owners and sellers may see things from a different perspective. There are many others differences:

- The life cycle
- The utilization cycle
- The operating environment
- The way an asset is operated
- The way an asset is maintained
- The view of when an asset should be renewed or over hauled
- The raw materials
- The expectations of the seller's role

◆ The expectations of the owner's role
◆ The seller's view of the owner's business model may be inaccurate.
◆ The owner's view of the seller's business model may be inaccurate.
◆ The owner and sellers view of their long-term relation ship as it pertains to the asset being exchanged

Although there are a lot of opportunities for misunderstanding, some of them may have an impact on the reliability of an asset; others may not. By highlighting the possible areas for misunderstanding, I am suggesting there is value in taking steps to eliminate those areas of misunderstanding. It may be impossible to make the sellers understand the owners' business needs or to convince them that the sellers are correct in their approach. But it is possible to ensure that the sellers understand the product requirements.

I am suggesting there is a value in using a process that will make those misunderstandings moot. As an owner, I only care about making sure the asset I am purchasing is provided in a manner that it can fulfill the requirements of my business model. Hopefully that will be consistent with the seller's needs. But if it is not consistent with the seller's needs I really don't care. After all, I'm the customer.

Identifying Requirements

The process of DFR for owners begins with a clear identification of their requirements. It concludes with a set of documents that clearly prove that those requirements have been met by the design (or that they are being met to the best knowledge and effort of the sellers). The two ends of the process should be clearcut. What is not so apparent is what comes in between.

Let's consider an assumption that is implicit to the process. This assumption is that the design will be iterative—there is no straight path from beginning to end. Once the initial concept for the asset design has been developed, the first step of the DFR process will be completed. This first design is one that fulfills the general requirements for functionality (e.g., production rate, product quality, product characteristics), but that is all. It is unclear how reliable, available, or maintainable the asset will be. After the first calculation of asset reliability is made, we can assume it will be necessary to change some element (component choice or configuration) to achieve the desired characteristics.

The same is true of the various characteristics viewed as a part of the DFR analysis. In order to understand availability and maintainability, it will be necessary to perform some form of analysis to identify all the tasks that will be required over the life of the asset. The identification of the tasks, their costs, their frequency, their duration, and their effectiveness (ability to extend the useful life or restore the inherent reliability) are all parts of the information needed to perform the availability and maintainability analysis. But when the results of those analytical steps identify a shortcoming when compared to the design requirements, it will be necessary to change the component selected or the configuration to deliver the required results.

The Iterative Process For Design and Analysis

Once the component or configuration is changed during the availability or maintainability steps, it will be necessary to go back and redo the reliability calculations (see Figure 4.1). When the reliability calculations are updated, the changes will have an impact on availability and maintainability analysis. They too must then be updated. By this time, the reader should be getting a clear picture of the need for integration and for the complexity and cumbersomeness that can be present if the systems being used to perform all aspects of the analysis are not tightly integrated. (Some suppliers have organizational structures that make this tight integration impossible. Owners should not have to suffer the consequences of the sellers' organizational weaknesses.)

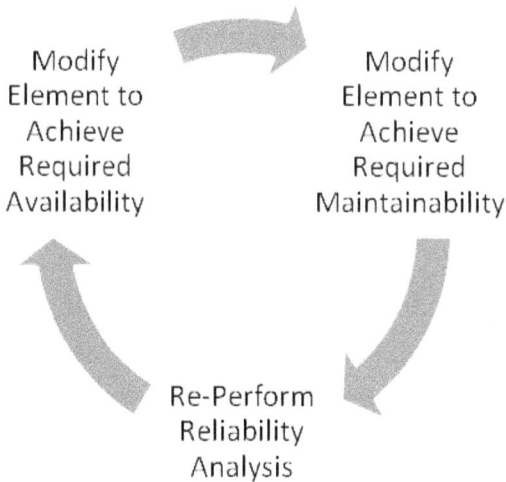

Modify
Element to
Achieve
Required
Availability

Modify
Element to
Achieve
Required
Maintainability

Re-Perform
Reliability
Analysis

Figure 4.1 The iterative process

Generally speaking, owners should reinforce the expectation that the iterative steps in the DFR process be completed;

they should ask the sellers to perform the analysis using systems that help streamline that integration. Start with an expectation that the sellers perform and provide the owners with the products of the following analysis:

1. Reliability Block Diagram analysis to calculate the expected asset reliability using data specific to the final configuration and the specific components that have been used in the final delivered product.
2. Reliability Centered Maintenance analysis to identify all the predictive maintenance (PdM), preventive maintenance (PM), replacement and repair maintenance, and overhauls or renewals needed to maintain the inherent reliability over its entire life.

A number of software developers provide software packages that include both of these analytical tools. The benefit of choosing a single source for both tools is that information can be shared. Performance simulations can be more easily completed if changes update both systems.

Assumptions about the Analysis
A basic assumption is that the analysis will require an RBD simulation and an RCM analysis. Another assumption is that the analysis will be far less complex if a software package that includes both tools is selected to perform the analysis. Inherent to both of these assumptions is a third—that the sellers have personnel capable of performing the analysis and using the selected software. Another implicit assumption is that the analysis is being done concurrently with the conventional design and, furthermore, that findings from the DFR analysis

are being integrated into the design.

It may seem obvious that the products of the DFR would be integrated into the design. Yet, there are many examples where they were not. My own experiences with project managers (including myself when I was one) suggest that they like to make decisions, then put those decisions behind them and move onto the next thing. Requirements—like DFR—that introduce numerous situations where choices need to be reconsidered, and apparently completed work must be reworked, go against the instincts of those project managers. As a result, they will push to put long-lead time components on order, even when their capabilities in terms of reliability, availability, or maintainability have not been verified.

It is important that all of the assumptions described above be verified as more than just assumptions. They need to be determined to be facts and a part of the process that will be used in the design of the new asset.

The Reliability Analysis step begins with a thorough understanding of how the remainder of the overall DFR will be conducted. It is critical to understand all design requirements as they relate to reliability, availability, and maintainability. It is important to understand how the Availability Analysis and Maintainability Analysis steps will be conducted. Reliability, Availability, and Maintainability analysis depend on each other for data. If not conducted in a closely-integrated manner, then additional, unnecessary work may be required. It is important to understand which of the available analytical tools (software) will be used. It is important to know as well that the sellers have the skills within their staff to conduct the analysis and to properly use the analytical tools that were selected. Without this knowledge, it is possible that the sellers could provide a lot of information and reports that are nothing more than "eye-wash." They will not ensure the required performance or help the owner maintain the desired performance over the life of the asset.

Steps of Reliability Analysis
Determine Owners Requirements

♦ **The life cycle**—Is the asset expected to last fifteen, twenty, or thirty years? Is it an accepted fact that the reliability and capacity will reduce as the asset ages or is it expected that the asset will have the same performance over its entire life?

♦ **The utilization cycle**—Is the asset expected to operate at full rates and in the harshest service for its entire life or will there be periods of lighter load or milder service?

♦ **The operating environment**—Where will the asset operate? What are the weather conditions? Will it be subject to salt spray from the sea? How about blowing snow and sub-zero conditions?

♦ **The way an asset is operated**—Are the operators well qualified? Will the asset be manned in a manner that operators will take good care of the asset? Or will the asset be used in a remote location where there is little or no human attention?

♦ **The way an asset is maintained**—Are the maintenance crews highly qualified or not? Is it likely that on-going maintenance and repairs will be done in a professional manner or will the work be done by "backyard mechanics"?

- **When an asset should be renewed or overhauled**—What interval is expected between overhauls? If the device has a thirty-year life, will it require two overhauls or three?
- **The raw materials**—If the asset is an operating plant, what are the properties of the raw materials it will process? If the asset includes an engine, what kinds of fuel and lubricants will be used? The importance of identifying raw materials is not simply to make a list. The importance is to understand the characteristics of the raw materials that will result in some adverse affect. In some cases, this is not obvious. An example is sulfur content in fuel. From an environmental standpoint, this change is good. From the standpoint of acid-gasses being produced, this is a positive change. From the standpoint of inherent lubricity, the change is not good.
- **The expectations of the seller's role**—Are the sellers expected to provide on-going and on-site service until all requirements have been met? If expectations have been met for some period, but change and no longer meet expectations, what is expected of the sellers?
- **The expectations of the owner's role**—The owners have some idea of the role they should have to play over the life of the asset. It may be that they simply push a start button, then sit back and let the cash roll in. If this is the case, the owners should be clear to the sellers so there is no confusion concerning how the asset should operate and the owners' role in achieving that level of operation.
- **The sellers' view of the owners' business model**—Sellers may view the owners' business model as containing an on-going income stream for the sellers (to provide parts and service for the life of the asset). If this is wrong, it is best to find

out early.

♦ **The owners' view of the sellers' business model**—Owners may believe that sellers have included sufficient cost in the initial price of the asset to provide on-going support for the remainder of the asset life. If this is wrong, it is best to find out early.

♦ **The owners' and sellers' view of their long-term relationship as it pertains to the asset**—Sellers may view the sale of an asset as being similar to a marriage in which the owners are tied to the sellers for the life of the asset. Owners might view the purchase of an asset as either a "one-night-stand" or the purchase of an indentured servant. In any case, it is best to know how long the relationship will last, what it will cost, and what services will be provided.

Complete Conventional Design—First Round

♦ Depending on the kind of asset and where the specific model or version fits in the on-going evolution of the generic form, the conventional design process can have a variety of differing steps. Considering the example of an automobile, the design process for a simple upgrade from the prior year model will be quite different from the design process needed to develop the company's first hybrid. Generally speaking there are a variety of specialized areas of design when starting from scratch, including:

♦ Functional design   ♦ Structural design
♦ Thermal design   ♦ Aesthetic design
♦ Design of control systems ♦ Etc.

♦ In order to allow the Reliability step of the DFR to proceed, it is important that the first round of the conventional design be substantially complete. It is necessary that the entire configuration be described (so it can be modeled) and the initial component choices are made (so their assumed reliability performance can be used).

Create Model

♦ If an RBD technique is used as a basis to calculate the reliability (and by now you should realize you are being strongly pushed in that direction), the next step is to assemble a model that closely resembles the function of the product. This step should be fairly easy for the individuals involved in the product design. If not, you should question how well they understand the product they are designing and how it functions.

♦ If sellers choose to use another approach (like a reliability allocation approach), it will be important to understand how they will address redundancy or the need for redundancy in the design. Another area of concern is the ability to evaluate where best to make improvements. RBD will describe specifically where changing a component will have greater or lesser impact. On the other hand, the mathematical impact of upgrading any of the components in a reliability allocation method is all the same. When using a reliability allocation method,changes can occur in subsystems where the current design is beyond the current allocation rather than where they are most needed or will be most cost effective. A final concern is the benefit of simulations in RBD. Simulations will allow failures to occur wherever they may.

Some years may have more failures and other years may have less.

Reliability allocation assumes failures are evenly spread. If failures are clustered in the first year of operation, owners may go broke before ever getting started. If real-life failures happen to be closely grouped in any later period, owners may not be able to meet production commitments for that period. This kind of problem is far more likely to be identified using simulations than strictly an allocation method. It is important for owners to be involved in the selection of the modeling approach that will be used and the software. These choices must fulfill the owners' requirements as well as the sellers'.

Apply Data

♦ Once the RBD or other model is complete, the next step is to apply the data needed to support the simulations. The data can come in a variety of sources but needs to accurately represent the anticipated performance of the device represented by each field. The most accurate form of information is probably failure data coming from a similar asset currently being operated by the owner. This data tends to blend in some reality based on how the owner currently operates and maintains the device. In some cases, the treatment provided by the owner can result in an extended life. In other cases, it can reduce the life when compared to the expectations of the supplier.

♦ In addition to the reliability data applied for each component, data is needed to describe the environment or system in which the asset operates. If the asset will be in a remote location and the owners do not plan to provide a local supply of spare parts, it is likely that the MTTR or Mean Time To Repair will be much greater than for owners who are either close to warehouses or have a substantial inventory of spare part.s

♦ As with other elements described herein, the selection and use of data must be agreed by the owners. There are occasions when data for the precise proposed element is not available. In those cases, it is critical for the owners to see that the data being used does not misrepresent or intentionally misrepresent the anticipated performance. The data should not cause the model to produce convenient results or results that seem to have been the objective before the model was constructed.

Produce Initial Results

♦ Assume that an RBD model is being used to provide simulations of the life of the proposed asset. For assets that have reliability measured in terms of mission success, each simulation will represent a typical mission. For most other assets, the simulation and calculated reliability will be based on some specific time period. In many cases, a year is the time interval being simulated. A year might be the best choice if it is consistent with annual service intervals or regular maintenance that can be most easily understood and budgeted when described on a yearly basis. Because most owners perform key forms of financial analysis on an annual basis, using an annual basis for reliability analysis will assist in determining how reliability related costs will affect financial performance.

♦ Most significant is the number of intervals or the entire life being analyzed. If the asset has a thirty-year life, then 30 one-year intervals will be used in the simulation. If the required life is shorter, a smaller number of intervals can be selected. The point is that the number of intervals and the total life is based on the owners' requirements and not those of the sellers.

♦ Another consideration is the number of simulations that should be completed. With the computing power of current computers, it makes sense to complete more simulations rather than fewer. The main concern is to complete a sufficient number of simulations to arrive at an acceptable confidence level. Start with at least one hundred simulations.

Conduct Availability and Maintainability Analysis

♦ Availability and Maintainability analysis will be discussed in greater detail in Chapters 5 and 6. The reason for mentioning them here is that the overall DFR process is an iterative process. It will be necessary to complete one or more cycles of availability and maintainability analysis before the final cycle or reliability analysis can be completed. Those analytical steps may result in changes in component selection or configuration that will ultimately affect the reliability analysis.

Iterate Conventional Design and DFR based on Findings

♦ As mentioned in the last bullet, if the designer is really trying to include reliability considerations in the design, the analysis is likely to result in a number of iterations in the design process. The initial conventional design will produce the first pass at a workable product. When performing the

DFR analysis, it is likely that some portion of the initial design will display one or more of the following problems:

- The overall asset will not provide the required level of performance. Therefore, one or more components will need to be changed or the asset configuration will need to be changed. It is possible that the model will drive the owner to make changes in the operating environment, like additional redundancy or shelf spares that reduce the expected down time.

- The overall asset does not provide the required level of availability so either components or configuration will need to be changed.

- Components or other elements of the current design do not have the required life or the required performance over the entire life.

- Components or other elements require more maintenance or replacement over the entire life and therefore have a higher Total Cost of Ownership than is acceptable.

- Some portion of the asset is not maintainable because it cannot be maintained in a ratable period of time. Furthermore, some portion of the asset is not maintain able because there is no assurance that the inherent re- liability will be restored using simple and clearly described maintenance steps.

♦ When any of these issues have been identified, the next step will be to change them in the conventional design to correct the observed weakness.

♦ Once the basic design has been changed and the new com- ponents or revised configuration has been integrated into the conventional design, it will be necessary to repeat the DFR steps to ensure the design requirements are being achieved.

Finalize Design

♦ Once the design for the basic functions and the required reliability, availability, and maintainability are complete, it will be time to finalize the design. This step is more than going out after work and taking a victory lap around the local saloon. It means that all the characteristics critical to both the functionality and the reliability performance are locked in. The concept of a design being "locked in" comes with one most-basic consideration: At this point in time, all changes are halted. The procurement department should stop looking for cheaper components. Programmers should stop tweaking software, adding features that may introduce defects. The customers should have stopped asking for changes sometime in the distant past. If they have not, this is certainly the time for halting owner-requested changes.

Create Report

♦ Referring specifically to the characteristic of reliability, it is now possible to produce the DFR Reliability Report. It is most important that this report be clear and usable by the owner. Sellers will frequently produce "proprietary" reports of various kinds of failure or risk analysis that are hundreds or thousands of pages long. Narrative reports of this kind are of little use to anyone, in particular the owners. The detailed data in this report should be provided in a database. Then the database can be quickly sorted and scanned for the specific data when needed.

♦ The first piece of information that is needed is the overall or macro-reliability of the overall asset by year (or discrete operating period) for the entire life of the asset. The report should show the number of reliability-related incidents during each period, the device that is expected to fail, and the failure mode that is expected. This information will be needed in the next part of the analysis to quantify the availability. It will identify the number and kind of events leading to the unplanned portion of unavailability.

♦ The second piece of information that is needed is the reliability for each component. This information should include three things:

• The reliability including a description of assumed failures during each period of the asset's life. This description will include both the failures that will cause an interruption of the entire asset and those that result in only a component failure.

• The maintenance required to ensure the inherent reliability of the component is preserved. Both predictive and preventive maintenance should be described.

• The timing at which it is expected this component will need to be replaced in order to maintain adequate performance.

Keep in mind; the DFR analysis for Owners is intended to do two things:

1. It is intended to provide a mechanism to ensure that sellers have performed an adequate level of DFR and that the analysis conducted has led to a product that meets the owners' needs.

2. It also provides a mechanism for the sellers and owners to share all the information pertaining to reliability in an efficient and effective manner.

Keeping the second item in mind, it is important for the sellers to tell the owners anything that may not be obvious from the reliability analysis. Probably the most important kind of information will be data pertaining to things we take for granted. My experience has been that subtle issues concerning re-sourced items—like piston rings or piston skirts or other components of that type—have caused the most unexpected problems because they were changed and the importance of the change was not recognized. When changes of these kinds occur and are not communicated directly from the sellers to the owners, the credibility of the seller can come into question. Based on personal experiences, the reader should be assured that all the excuses in the world do little to restore the owner's confidence.

Conclusion

I hope that by now you clearly understand the level of owner engagement in the design process needed to achieve the desired results. If owners choose to remain aloof and rely on sellers to provide a reliable asset, they will receive an asset that meets the seller's requirements and definition of reliability. For the asset to meet their own requirements, owners must be deeply involved through every step of the design process.

Chapter 5

Availability Analysis

*Few things help an individual more
than to place responsibility upon him,
and to let him know that you trust him.*

Booker T. Washington

Availability analysis can be performed in a variety of ways. Some of them are closely linked to the manner in which the Reliability analysis is performed. In fact, some computer programs used to perform Reliability analysis also provide an estimate of the anticipated availability, based on the unavailability associated with events caused by poor reliability. This chapter focuses on a way to analyze Availability that will be meaningful to the long-term owner.

Although the software programs described above provide a starting point, they ignore both planned unavailability and many of the logistical and real life restraints that cause the actual unavailability to be far greater than the number that is calculated. Suppose your car breaks down at a place where servicing and parts are readily available. The resulting event will cause far less of an interruption than if it breaks down at some location where neither are readily available. We can see, then, that Availability analysis must clearly represent the environment in which the owner operates. It must also present the results in a manner that the owner can understand and can enforce if not meet.

Defining Availability

Availability is defined as the portion of time that a system or plant is able to perform its intended function. Conversely, Unavailability is the portion of time that an asset is not able to perform all the functions deemed critical. For example, the critical functions of a car are those pertaining to its ability to provide transportation and specifically its ability to provide safe transportation. Clearly that function includes a car's ability to move passengers to their desired destination and to do so safely. The important systems include those that provide motive power, steering, and braking.

On the other hand, most cars are equipped with radios for entertainment. However, this function is not considered a critical one; if the radio stops working, the asset has not suffered a loss of availability. There are other systems that do not prevent the car from providing transportation, but may be viewed as borderline functions. Exterior lighting is one of these systems. You may get a warning ticket if head lights or brake lights are not functional, but would their failure prevent you from using the vehicle? Although you would have them repaired as soon as conveniently possible, few people would park their car until the lights are replaced. You would continue to use your vehicle, even if the heating and air conditioning systems are not functional. But what if your sole use is for trips through the desert, or through wintery conditions, and you need to transport an elderly, infirm passenger or a new baby? In these cases, the functions defined as being critical may be different from other cases.

It is obvious that an asset cannot perform its function if it is completely shut down. There may be other periods as well that it is unable to perform its intended functions. For instance, if the function is to produce a product of a specific quality, but the system is unable to meet that quality, the asset may not be performing its function despite the fact it is operating and producing product.

Calculating Availability

The following equation can be used to calculate Availability:

$$\text{Availability} = \frac{\text{Total Time in Operating Cycle} - \text{Downtime}}{\text{Total Time in Operating Cycle}} \times 100\%$$

In this calculation, I use the term Total Time in Operating Cycle to refer to the maximum period covered by any non-repetitive event. Suppose every ten years an overhaul is needed that is somewhat longer than any other maintenance activity. (It may be to allow for an in-depth inspection or repair that requires more extensive disassembly than on any other occasion.) The Operating cycle would be ten years or 3,650 days. Selecting a cycle duration that includes all forms of maintenance event, specifically the longest, ensures that the calculated availability is representative of the overall life cycle.

The equation above applies to assets that are required to provide their functions on a continuous basis. If the asset provides support for a series of specific missions (like a space shuttle), the calculation would be based on the asset's readiness to complete the mission whenever scheduled. For example, a person's car might be required to fulfill a series of discrete missions. Nevertheless, most cars are expected to be available to transport the owner 24-hours per day and seven days per week, whenever and wherever the owner wants to go. As a result, a failure is not simply viewed as the number of times the vehicle fails during a trip. A failure is also any occasion when the vehicle is not ready to perform when called upon.

Calculating Downtime

In calculating the downtime, two categories should be considered. One is unscheduled downtime resulting from unanticipated reliability related events. The other is scheduled downtime resulting from scheduled or planned outages. In determining the period of downtime for both categories, all aspects must be considered including time for operating steps, logistics, and any other cause of delays.

Ultimately, the calculated or modeled reliability tells the number of failure events or the likelihood of a specific failure event that can be expected during any given period. In addition to knowing how many events occur, it will be necessary to estimate how long will be required to recover from each event. If an RBD analysis has been used to determine the reliability, the statistical reliability of each component or equipment item will provide an idea of which component is most likely to fail. Once it is known which items are most likely to fail, it is necessary to apply first-hand experiences to estimate how long will be required to make repairs and return the asset to full production. This subtlety is important. The unavailability resulting from a specific event may vary widely from location to location and from owner to owner. If an asset is in a remote location and the owner does not choose to maintain the resources needed for a quick recovery, the availability impact might be far greater than for an unexpected interruption in a "convenient" setting.

For instance, if the overall reliability of a facility is 90%, there is a 10% likelihood of failure in any year. If the operating cycle is ten years, there is likely to be one failure event in each operating cycle. Suppose it is determined that the most vulnerable component is the major recycle compressor and experience says it takes 30 days to recover from a failed compressor. You can then estimate the unscheduled loss due to poor reliability at 30 days in every 3,650 days. This calculates to .82% unscheduled unreliability.

For other assets (like locomotives or heavy mobile equipment), the owners may choose to specify the maximum or allowable unreliability. They may desire calculations that provide data in terms such as FLY (Failures per Locomotive Year) or the number of failures that each asset in a fleet will experience in a complete one-year period. For instance, a requirement of "One FLY" performance means that a maximum of one failure (or loss of critical function) is allowed for each locomotive in

a fleet per year. If there are fifty locomotives in a fleet and the fleet is made up of "One FLY" locomotives, the owner should expect 50 failures per year or nominally one per week.

Once the anticipated frequency is known, and the typical outage duration is determined based on the most likely form of failure, it will be possible to calculate the likely unavailability resulting from unplanned or unscheduled causes. For instance, reviewing all the possible unreliability events, if the average "time to recovery" is five days, the unavailability for the case listed above would be five days per year for each item.

Run Limiters and Duration Setters

The next area to address is the amount of time lost due to scheduled outages. The total impact of a scheduled outage should be identifiable well in advance. Two characteristics of any complex physical system are useful in identifying the frequency and impact of each scheduled outage on system availability. They are:

- ♦ Run Limiters—A specific component or circumstance that limits the maximum interval between outages.
- ♦ Duration Setters—The specific component or condition that sets the minimum critical path duration for a specific outage.

Run-limiters can be regulatory requirements or system limitations. An example of a regulatory requirement is a state regulation that mandates internal boiler inspections. Some states require annual internal inspections. Some states allow intervals as long as four years between internal inspections with external inspections during intervening years. An example of a system limitation is catalyst life. Some forms of fixed bed catalyst will remain effective a maximum of one or two years. In these cases, the catalyst must be replaced or regenerated on the maximum interval or suffer from reduced effectiveness. The following are other examples of issues that can create run-limiters:

♦ Corporate requirements based on age of unit or other issues

♦ Half-Life of pressure retaining equipment wall thickness

♦ Wear and tear resulting in reduced performance for key equipment

♦ Need for cleaning because of contamination build-up (e.g., fouling of heat exchangers)

Regulatory requirements to renew operating certificates may seem subjective and unnecessarily costly. Nonetheless, they establish the maximum interval between outages for elements like boilers that have been the source of catastrophic failures at some time in the past. It is best not to ignore the knowledge that was learned by the painful experiences of others and have to relearn it hrough first-hand experiences.

Duration-setters are those components that result in the longest sequence of tasks that must be accomplished once the asset is shut down. For operating plants, the string of events can include shutdown, isolation, decontamination, inspection, repair, re-commissioning, isolation removal, and start-up. Although there might be some exceptions, an accepted practice has been to use the current critical path duration for similar equipment as a surrogate for the expected duration for new plants of a similar design. Owners who do not have equipment similar to that being designed will have to depend on the knowledge of the designer, benchmarking of companies with similar facilities, or a high-level critical path plan.

A simple way to identify the run-limiters and duration-setters for any facility is to create a spreadsheet showing the following characteristics for each major sub-system in the unit (see Figure 5.1).

- ◆ Sub-system name
- ◆ Run-limiter
- ◆ Maximum life
- ◆ Duration-setter
- ◆ Minimum outage duration

After compiling the required information for each and every major sub-system, the last step is to select the Run-limiter with the shortest life and the Duration-setter with the longest duration. The interval between outages and the outage duration determined by these two items will set the Availability based on scheduled outages.

Figure 5.1 shows a System XYZ that contains three sub-systems. Note that the maximum interval between outages is determined by Component 2 in Sub-system B. The duration of the outage is determined by Component 4 in Sub-system A.

If a specific plant requires outages of different kinds for different reasons, it will be necessary to identify the frequency and

| System XYZ | | | | | | |
Sub-System	Component	Maximum Run-Length (Years)	Sub-System Run-Limiter	System Maximum Life (Years)	Duration Needed to Maintain (Days)	Minumum Outage Duration (Days)
A	1	3.5	3.5		15	
A	2	4			20	
A	3	4.5			17	
A	4	5			25	25
B	1	3.7			22	
B	2	3.2	3.2	3.2	23	
B	3	4			12	
C	1	4	4		15	
C	2	5			17	
C	3	4	4		20	
C	4	6			18	

Figure 5.1 Working with run-limiters and duration-setters

outage duration of each kind. For instance, if a plant depends on a steam boiler for operation and processes a corrosive material, there might be annual outages to address the boiler inspections and less frequent outages to address deterioration resulting from corrosion. Note that a separate spreadsheet like the one described in Figure 5.1 will be needed for each kind of outage the asset must experience. The interval between outages and the maintenance of all components that must be addressed during each kind of outage will determine the timing and duration of each outage.

In this situation, the interval between boiler outages is known. The duration of boiler inspection outages is simply the critical path duration of the steps needed to access boiler internals and perform the annual inspection.

The interval between corrosion-related outages is not so simple. Companies having either internal rules or abiding to American Petroleum Institute standards may conduct outages based on the "half-life" of the device where combined effects of corrosion rate and remaining corrosion allowance result in the shortest interval. (The half-life is the amount of time needed to corrode halfway through the corrosion allowance.) A significant amount of analysis is needed to understand these issues and calculate the anticipated interval between outages. In this case, the outage duration is not easy to determine because it depends on the amount of time needed to perform the inspection, take metal thickness readings, determine remaining life, and take corrective action if needed. If the remaining half-life is less than the amount needed to allow the desired run length, it will be necessary to replace the corroded component or replace corroded areas with weld metal.

This may sound like a lot of analysis to be accomplished before the asset is constructed. However, if it is not done, it will be impossible to:

1. Determine the actual expected availability.
2. Identify which components should be built in a robust manner or from a higher alloy material.

Although much of the attention thus far has been focused on operating plants, the same kind of philosophy applies to other kinds of assets like mobile equipment for instance locomotives. In the case of a locomotive, there is an obvious need for a major overhaul event every eight-to-eleven years. In addition, modern locomotives may require a "maintenance event" on a more frequent interval to address wearing elements like switchgear. These components need to be maintained on a regular basis, but would extend a routine maintenance interval by too much if crammed into a normal maintenance event. As a result, extended maintenance events may be planned on intervals between overhauls. The Availability Analysis needs to account for all these forms of outage.

Make It Easy to See and Understand

After the tasks of identifying the frequency and duration of all the outages have been completed, it will be possible to assemble a bar chart showing the nominal timing and down time during each outage. This chart can then be used to determine the number of days the unit will be down during any complete operating cycle. (Before the conclusion of this chapter, we we will also show how this approach should be used to describe the availability over the entire useful life of the asset.)

As an example, let's assume:
1. The unit contains a single boiler that must be inspected every year, resulting in a seven-day outage.
2. The unit requires a limited outage for catalyst change and exchanger cleaning every five years that entails 30-days downtime.
3. The unit requires a more extensive outage every ten years for replacement of corroded components that entails 45-days downtime.

The bar chart in Figure 5.2 represents the total downtime for each year in the ten-year operating cycle. As long as the maintenance intervals or durations do not change, this chart can be used on a repetitive basis to represent the planned outage time.

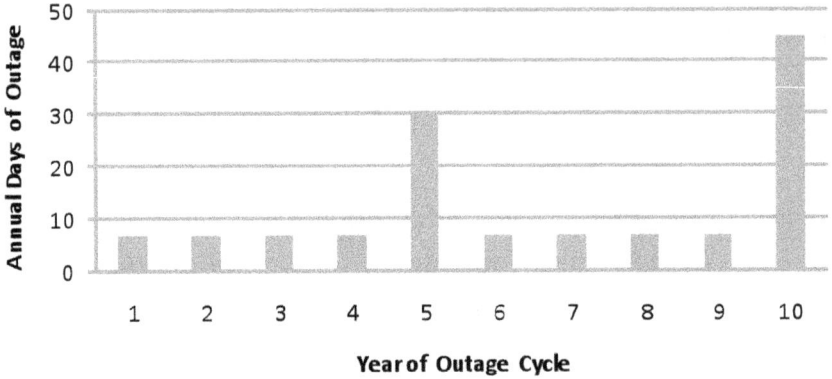

Figure 5.2 Outage Days

In total there will be 131 days of scheduled downtime every ten years. The "scheduled" availability will be:

$$A = 3650 - 131 / 3650 = 96.4\%$$

Keep in mind there is also an Availability debit of .82% due to reliability (rounded to .8%). The anticipated Availability due to both scheduled and unscheduled causes is:

$$Availability = 96.4\% - .8\% = 95.6\%$$

If the owners had specified an Availability greater than this number, they will be disappointed. As with the RBD analysis described in the last chapter, performing this analysis during the design process gives the owners the opportunity to require changes before the design is complete and the plant is built. In the case of Availability, the designer can improve Availability by:

1. Selecting a more reliable design (reducing unscheduled outages).

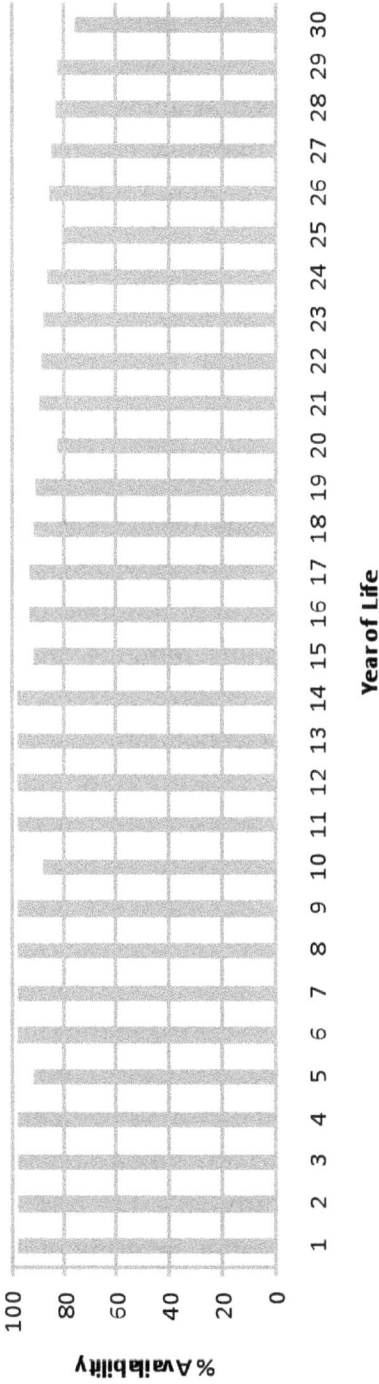

Figure 5.3   Percent Availability without Restorative Measures

2. Selecting Run-limiters that will allow longer runs between outages.
3. Selecting Duration-setters that allow for shorter outages.
4. Creating a configuration that may completely eliminate some of the outages (like a spare boiler in the example).

Several events might happen over the life of an asset that will adversely affect its Availability. For instance, several years ago, OSHA 1910.119 was placed into effect; it created a new "lock-out-tag-out" requirement that increased the amount of time required to isolate equipment during outages. This kind of change can increase the unavailability. When events like this occur, it is important for planners and schedulers to get back to work finding ways to reduce the time required by other activities so the overall unavailability remains the same or is improved.

Many assets have an expected life of twenty or thirty years. If left unaddressed, the reliability and availability will degrade over time. Therefore, it is important for the sellers to provide a graph that portrays the availability (and unavailability) over the entire life. If it is expected that some greater investment in "out" years will be required than in early years, the sellers should describe where the investment will be required and how much it will be (in present day dollars).

For instance, Figure 5.3 shows a sample availability bar chart for a thirty-year life if restorative measures are not taken.

On the other hand, the thirty-year life may look life the chart in Figure 5.4 if some restorative investment is made at the fifteen year point.

If investment is required to keep cash flow at the expected

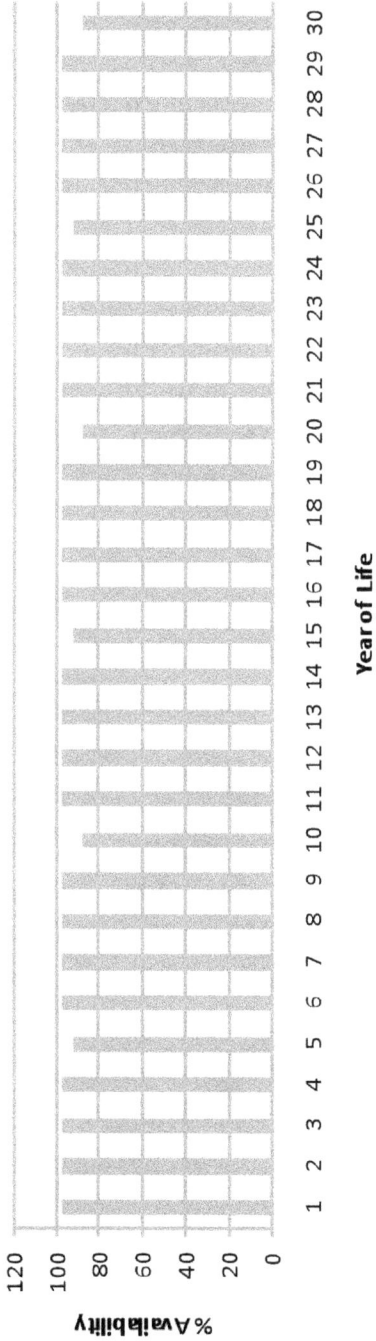

Figure 5.4 Percent Availability with Restorative Measures

rate, the sellers should tell the owners so they can build this information into their business model. The owners will either need to make the investment or accept the reduced production rate resulting from increased unavailability.

The name of the game is Return on Investment. Owners need to begin operations with a good idea of the amount of return they have a right to expect for their investment. The sellers are in a much better position to understand the issues that are likely to have a negative impact on this issue than are the owners. After all, the sellers are in the business of making and selling the asset, whereas the long-term owner is in the business of using the device to produce some other product for which they are the expert.

Conclusion

It is critical that the owners be able to rely on the sellers' knowledge of their product and of each and every component that goes into building their product to determine the expected availability. When owners find they cannot depend on the sellers to provide this knowledge, the owners typically decide they must become "experts" for the sellers' product as well as their own. When that happens, sellers suddenly find owners knowing more about their products than they do. This position is an uncomfortable one to be in. It is good for neither the sellers nor the owners. In this situation, sellers find themselves struggling to stay a few steps ahead of the owners' knowledge of their product, its weaknesses, and the source of the components from which it was built. Owners who find themselves in the position of being unable to obtain needed information from the sellers must spend lots of extra time and resources dealing with the problem.

It is best for owners to know and specify their requirements to the sellers and, in turn, for the sellers to provide the engineering and information needed to assure that those requirements are met.

Chapter 6

Maintainability Analysis

Management means, in the last analysis,
the substitution of thought for brawn and muscle,
of knowledge for folkways and superstition,
and of cooperation for force.

Peter F. Drucker

What Is Maintainability?

Maintainability measures the ability to restore the Inherent Reliability in a ratable period of time.

Inherent Reliability

The first of two abstract concepts that are used in the definition of maintainability is Inherent Reliability. The inherent reliability of any device or system is the highest level of reliability that can be achieved based purely on the configuration and robustness of included components. If a system is well designed, has redundancy where needed, and uses robust components, it is likely to have good inherent reliability. (A robust component is one that performs its intended function while surviving extremes of environment and misuse.) To achieve full inherent reliability, it will be necessary to apply the best possible operations and the best possible maintenance. Poor operations or maintenance will result in a level of performance at something less than the full inherent reliability.

Ratable Period of Time

The second abstract concept is that of a ratable period. A ratable repair is simply one that can be accomplished in a known, well-defined period of time. If the sequence of events needed to perform the repair is full of surprises or steps with unclear durations, the maintenance activity is not ratable.

Suppose you take your car to a mechanic, who says, "I can have your car done in two hours, but I don't know how long it will last." Your car is not maintainable. If the mechanic says, "I don't know how long it will take, but when I get done it will be as good as new," it is again not maintainable. To be maintainable, the mechanic will need to be able to say, "I can complete the work in X hours and it will be as reliable as new."

Identifying Maintenance Needs for the Entire Life Cycle

The process of evaluating the maintainability of any system

or plant begins with knowing what maintenance will need to be accomplished over its entire lifetime. There are several ways to determine the specific tasks that will need to be done over the life of an asset. One is to review the list of tasks that are being done to maintain similar assets that currently exist. Access to an asset with similar needs tends to make the maintainability analysis fairly simple.

On the other hand, let's assume that the asset being considered is either new or such a departure from earlier assets that past experience is not meaningful. In this case, the best and most thorough way to approach the chore of identifying all the required tasks is to perform a streamlined Reliability Centered Maintenance (RCM) analysis.

The seller could choose to perform a conventional RCM analysis rather than a streamlined method, but it is more time consuming and resource intensive. As a result, it is less likely to be completed in a timely manner. I understand that there is an on-going battle between those people who feel that streamlined RCM is akin to blasphemy and should never be used and those who think streamlined RCM is fine. In this case, I will call upon my experiences in attempting to convince organizations to perform any form of RCM. The argument against performing RCM is that it consumes too many resources and does not produce an immediate return on investment. If RCM requires too many resources and takes too long, it simply will not be done. Because som form of analysis needs to be done to produce a reasonably accurate and complete list of tasks, I recommend using a streamlined form of RCM for this application. The streamlined forms of RCM will produce acceptable results and can be done with the least amount of resources in the shortest time.

Proactive and Reactive Tasks

When the RCM analysis is complete, the finished product will be a list of both proactive and reactive tasks that will occur on a regular basis over the life of the asset. The proactive tasks (PM/PdM) will be identified directly. The reactive tasks will be identified by default. When the economic analysis used as a part of RCM identifies a "run-to-failure" situation, by default, it is identifying a reactive or repair task that will be required during the life of the asset.

Using the complete list of tasks produced by the RCM analysis, the reliability engineer involved in performing DFR during the design of the asset should perform a mental (and if possible physical) walk-through of each task. The walk-through needs to answer two questions:

1. Is it possible to perform this task in a manner that will restore the Inherent Reliability the first time and every time thereafter?
2. Can this task be done in a ratable period of time?

First, consider examples that might lead one to answer "no" to the first question:

- The task requires diagnostics of an electronic device (like a computer board) and there is no way of testing the device.
- The task requires some form of connection or fastening that is impossible to perform in the field in the same manner it was during manufacture.
- The task envisions cleaning that cannot possibly remove all the debris.

♦ The task envisions addressing deterioration in a manner that cannot possible renew the component to like new condition.

In order to ensure that the asset is maintainable, it will be necessary to find a way to make a "yes" from each "no" that is identified.

Now consider examples that might lead one to answer ?no? to the second question:

♦ The access panel is too small.
♦ The craft performing the task will require scaffolding or fall protection where none readily exists.
♦ The repair requires soldering or welding in an awkward position or by individuals who are generally not quali fied.
♦ The task requires the person performing the work to be in an awkward position or confined area.

Again, the list of problems that can make the time required to complete a task unsure are too numerous to list. As before, it will be necessary to find a way to make it a "yes" from each "no" that is identified.

Turning No to Yes

Rather than attempting to pro-vide a comprehensive list of ways to turn each "no" into a "yes," I will provide a single example that covers a variety of issues.

Military assets used in difficult environments frequently

depend on exchange of complete modules to ensure that "like-new" reliability is restored in a ratable period. An example is the communication radio on war ships. When signs of incipient failure are detected, the radio module is replaced with a "Hot-running-spare" that is known to be functioning perfectly.

This military example may apply to relatively few instances in private industry, yet it serves as a useful example of what is possible. Despite any arguments about the costs or applicability of having hot-running-spares available, it is still important to ensure that mechanics do not need to stand on their heads to complete the needed repairs.

Conclusion

In my own experience, "maintainability reviews" began to be conducted 20–25 years ago. When the concept was first introduced, no one knew how to perform one. Typically, the review consisted of making sure that the most remote or difficult crane lift could be accomplished by the largest crane in the plant. Another consideration was that the smallest manway in a new plant was able to accommodate the largest craftsperson. (I recall having to use Crisco shortening to help retrieve a particularly large person.) Ultimately, those early reviews did very little good.

Plants that were reviewed as described in the last paragraph ultimately continued to suffer from poor reliability and poor availability. If the objective is good reliability, the approach used to conduct maintenance must also restore the inherent reliability. If the objective is good availability, the approach used to conduct maintenance must be handled in a well-controlled manner for which the time to repair is known.

Although the process for ensuring maintainability described in this chapter is more cumbersome than the old approach, it will certainly deliver better results.

Chapter 7

Organizing for Concurrent Engineering

*If you want to build a ship, don't herd people together
to collect wood and don't assign them tasks and work,
but rather teach them to long for the
endless immensity of the sea.*

Antonio De Saint-Exupery

The first time I tried to introduce DFR as a part of the design process for a new plant, I arranged a meeting with the Project Manager to explain the concept to him. At the conclusion of that meeting, he said it sounded like a good idea, but that for the current project, I was too early in the design process and should come back later. I returned after a few weeks and was told that I was too late. He said it was unfortunate that I was too late, but that with the current staffing, the introduction of an additional step would result in rework and unacceptable delays.

As you might expect, the application of DFR to the design process is neither simple nor easy. The DFR process introduces new steps to the design process and frequently causes changes to be made. One characteristic that seems to be common to most Project Managers is that they do not like change.

Risky Investments Without Assurance of Reliability

Nonetheless, companies that are investing hundreds of millions—or, in some cases, billions of dollars—in new assets are increasingly demanding to be assured they will operate in a reliable manner when complete.

Much smaller companies are at even greater risk. Many small interests who build renewable energy facilities of one kind or another are really in the Return On Investment business. In other words, they assemble a certain amount of capital from their investors with the expectation that the capital will provide investors with a reliable income each year. These small companies choose to invest in an ethanol plant, bio-diesel plant, or wind farm, not only because of their reliable income, but also because they provide a way for investors to support their own community.

When these small interests are "sold a bill of goods" concerning what they have a right to expect from their investment, the impact is far more dramatic than with a large company that may have other sources of cash to fall back on. It is critical for these small interests and the investors who support them to receive exactly what they expect. As a result, the design process used to develop the equipment they purchase must address all the elements needed to provide a sound and reliable product. These small business owners need to know what their assets will produce and how much it will cost to keep those assets producing.

This chapter begins by emphasizing the diametrically opposed positions between the need of the customers and the willingness of the design agency or the sellers to meet that need. My objective is to show you clearly what you are confronting when you begin demanding the analysis and documentation needed to ensure reliability. On one side, reliability is the single issue that causes many businesses to either survive or fail. On the other side, organizations involved with the development, design, and sale of assets are often unwilling to change the way they do things to address the needs of their customers. When discussing the subject of this chapter—creating the organization needed to conduct DFR—the first challenge is getting past this disagreement. DFR requires resources. Furthermore, those resources need to be in the right place at the right time to make a difference.

Making DFR a Required Part of the Design

The three characteristics that are critical in determining how the new assets will perform over their entire life span are Reliability, Availability, and Maintainability. As a result, owners are increasingly asking for these characteristics to be addressed during the design stage of a new asset. If the owners have some experience with the steps needed to ensure the delivery of these requirements, they include a requirement for DFR in their design specifications.

Most design specifications already contain some wording that can be interpreted as requiring that the product contain some semblance of these three characteristics. If the specifications require some amount of robustness be built into the asset, that robustness will be accompanied with a certain ability to survive. If the specifications call for fasteners that meet SAE standards or for materials that meet certain ASTM standards, those requirements will accompany some assurance that the inherent quality of components meets some minimum standard.

Yet, those forms of specifications do very little to ensure that either the overall asset or the components from which it is made are capable of specific performance levels in the areas of reliability, availability, and maintainability. If they do, it is more a matter of luck than science or engineering. To ensure that reliability, availability, and maintainability issues are addressed in a scientific manner using engineering techniques during the design, it is mandatory that individuals having the skills to perform DFR be involved during all phases of the design process.

It would be a mistake to assume that the same personnel who perform conventional design activities can or will perform the DFR analysis. Although individuals assigned to perform the process design, the hydraulic design, the thermal design, the control system design, or the structural design are knowledgeable of their aspects of the design process, they typically have neither the experience with DFR nor the time to perform DFR when it needs to occur during the design process. As a result, a separate group of individuals need to be assigned and held accountable for the reliability design.

Concurrent Engineering

Concurrent engineering is another term used to describe the process of addressing reliability aspects during a system design. This term is particularly well suited to describe the DFR activities during the design of a new system. To be effective, DFR must occur at the same time or concurrently as other aspects of the design. The term is also useful because it tends to emphasize the importance of the timing of DFR analysis. It is difficult for stubborn Project Managers to say they thought it could be done at any time if, from the very start, they are told it is a concurrent activity.

DFR cannot be left until after the plant is laid-out, with piping designed and equipment selected. It needs to occur while the initial design is in process. Changes can then be made when it is still cost effective and time-efficient to do so. If too much progress is made to the conventional design (e.g., equipment locations are determined, piping is laid out, equipment is selected), it will be difficult and costly to change. Conventional design steps and DFR are accomplished in an iterative manner—the conventional design offers initial concepts and DFR determines if those concepts are consistent with the Reliability, Availability, and Maintainability requirements that have been set by the owner. Reliability, Availability, and Maintainability must be confirmed before the design can be "cast in concrete."

It is critical that the individuals assigned to perform the specific elements of DFR be onboard and part of the design team from the very start of the design process. Their participation is important for avoiding the "too early" or "too late" quandary mentioned at the very start of this chapter. Reliability, Availability, and Maintainability performance requirements need to be clearly identified as part of the design requirements. The resources specifically accountable for delivering those requirements must be identified at the same time.

Making the Case for Added Resources

Let's begin by tackling the argument that many Project Managers might have when told that the reliability personnel need to come on-board at the same time as the rest of the engineering team. The Project Manager might argue, "If they come on board from the very beginning, they won't have anything to work on. It will be a waste of money until the engineering team has produced something for them to analyze."

It turns out that very few things being designed are completely brand new. Most products are an extension of something that preceded it. The product is simply larger (or smaller), or performs its functions in a slightly different way or adds a new function. Therefore, the opportunity to begin performing Change Point Analysis is introduced as soon as the decision is made to begin the development of the new asset.

The current asset has some quantifiable performance level in the areas of reliability, availability, and maintainability. The performance in those areas either meets the new requirements or it does not. If not, work can begin on identifying what needs to change to meet the new requirements. In addition, the new product has a specific list of changes from the current asset (like the current model of sensor being used is no longer made). For all the changes, it will be necessary to identify the elements of design that are critical to achieving the required performance.

It is naïve to believe that little can be done from the start of a design to ensure adequate reliability performance. History is a good teacher when it comes to this subject. Even products that are part of the evolution of a long line of very reliable products often turn out to have their own reliability problems because subtle changes have been missed.

Designers can miss the point that changing or improving a simple and reliable design may introduce stresses that did not exist in earlier models. As products evolve, suppliers may choose to make their own improvements; some of these may improve one aspect of their manufacturing process while harming others. Changing from mechanical to electrical and then to electronic to microprocessor-based controls may seem to be obvious improvements that go with the evolution of technology. But if these changes are not adequately engineered, problems can result.

This initial change point analysis by the reliability engineers will ultimately be useful for other members of the engineering team. The analysis will help focus their attention on areas of potential risk and opportunities for improvement.

Once the other members of the engineering team have begun to make progress—either developing the configuration of the new product or selecting candidate components to fill out the gaps in the system—the reliability engineers can begin modeling the proposed systems or evaluating the reliability performance of proposed components. The integrated design team and design for reliability team needs to make every effort to avoid the traps inherent to many fast-track projects. For example, they must see that all analysis is complete before equipment and components are placed on order. It is difficult and costly to back away from long lead time orders once they have been placed. This is particularly true when the equipment currently on order will fulfill functional requirements and it is "only" the reliability-related requirements that will not be met. Anyone who has been in the reliability engineering business for any substantial period of time can point to a number of examples where just that choice needed to be made and it was made in favor of expediency over reliability.

Another question concerning the design organization has to do with just how many reliability engineers are needed to keep up with the design team without becoming the excuse for delays. Once the need for a reliability engineering presence on the design team is recognized, and the other design team members begin to understand the function of the reliability engineer, it will become more apparent:

- ♦ What needs to be done
- ♦ When it needs to be done
- ♦ How many resources are need ed to perform the reliability engineering function in a timely manner

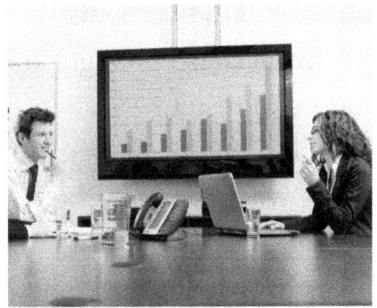

Very large projects need multiple individuals performing the reliability analysis function. Projects that are dependent on highly-specialized types of equipment will benefit from having reliability engineers with specific experience using that kind of equipment. Suppose your project includes a world-scale, high-speed centrifugal compressor. It would make sense to have a reliability engineer experienced with those kinds of machines to work with the rotating equipment engineer who is selecting it. Although some folks may say that the rotating equipment engineer should be able to adequately perform the reliability analysis, I have found that not to be the case. Many specialists fail to view their equipment from a systemic standpoint. For instance, a pump may be robust in and of itself, but when operating as part of a complex system, the "pumping system" may be inadequate. Someone must ensure the overall system is reliable. That someone is most frequently a reliability engineer who is accountable for the overall system performance.

On one occasion, the rotating equipment engineer specified a system with far more redundancy that was justified by the value of the risk of failure. On a second project of the same kind and size of plant, the reliability engineer's analysis was able to reduce the cost of a specific system by nominally one-third. In almost every situation, the project will benefit from the application of the various elements of the DFR process. On some occasions, the timing and the application may seem painful, but that is only for lack of familiarity and understanding.

Using the System Engineering V-Model to Reduce Resource Requirements

Although the reliability and long-term performance of capital intensive assets are critical to every owner, not all owners have the same amount of resources or the same ability to address this important issue. The most effective approach might be to devote full-time and fully qualified assets to be "embedded" on-site with the seller's designers and builders for the entire duration of the development process. However, that is not always possible. As a result, it is important to identify alternative ways to achieve adequate results.

Once again, the system engineering V-model provides a useful tool in understanding how best to leverage limited resources to achieve the best results. Figure 7.1 provides some basic suggestions of ways an owner can obtain the greatest impact with limited resources. Appropriate involvement at key points in the design process can help identify when characteristics that will determine the system reliability, availability, or maintainability are getting off-track.

As was discussed earlier, the individuals pursuing improved reliability need to understand clearly what is required and how the requirements are to be achieved at each level. With this approach, owners must build an expectation that the sellers will provide proof of conformance and act in a very open and forthcoming manner. As with any trust-based approach, the assumption is that the entity being trusted will act in a way that shows it is deserving of the trust. Furthermore, the entity that is trusting must perform enough due diligence to assure the trust is deserved. Any circumstance where trust is found to be violated must necessarily result in some penalty.

As shown on the V-model in Figure 7.1, owners should begin the oversight process by agreeing to a schedule of regular meetings with the sellers. These meetings should begin with a presentation by the sellers of a detailed description of their design process and the schedule on which it will be completed. This step might entail several iterations until the owners arrive at the point they feel they understand the process and schedule in adequate detail to ensure they can be "at the right place at the right time" for all the key milestones through the design-and-build process. The key difference between this approach and other more resource-intensive approaches is that individuals providing oversight are involved only a small part of the time but at critical times. In other approaches that involve more

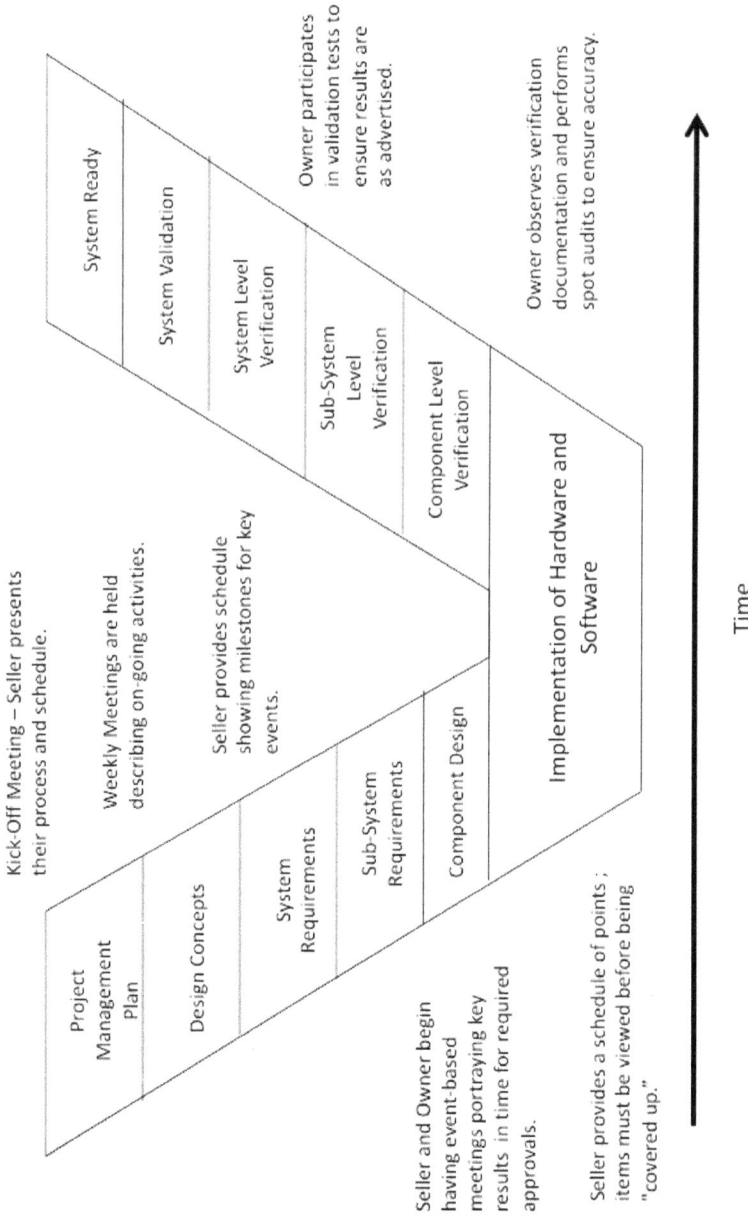

The diagram (rotated) shows the System Engineering V-Model with the following labeled elements:

Left/descending side:
- Project Management Plan
- Design Concepts
- System Requirements
- Sub-System Requirements
- Component Design

Bottom:
- Implementation of Hardware and Software

Right/ascending side:
- System Ready
- System Validation
- System Level Verification
- Sub-System Level Verification
- Component Level Verification

Annotations:
- Kick-Off Meeting – Seller presents their process and schedule.
- Weekly Meetings are held describing on-going activities.
- Seller provides schedule showing milestones for key events.
- Owner participates in validation tests to ensure results are as advertised.
- Owner observes verification documentation and performs spot audits to ensure accuracy.
- Seller and Owner begin having event-based meetings portraying key results in time for required approvals.
- Seller provides a schedule of points ; items must be viewed before being "covered up."

Time

Figure 7.1 The System Engineering V-Model.

resources, the individuals performing oversight are involved at the right times because they are there all of the time.

The following points describe the owners' part time involvement in greater detail:

- Start by clearly understanding the sellers' design and development process so you can plan your schedule for oversight. Tell the sellers your plans so that they know that intentionally deviating from the schedule is unacceptable.

- Begin by holding regular meetings. Although they may seem cumbersome and time consuming, the routine will be useful in providing an opportunity for identifying all issues and milestones.

- Once the project and requirements have become clearer to both the sellers and the owners, update the design development schedule with more accurate events and dates.

- Begin holding event-based meetings. Rather than meeting weekly or bi-weekly, meet whenever needed. To reduce the resource requirements, use teleconferences and web-based meetings for updates. Early on in the process, follow up with an on-site audit to verify that everything is as expected. If there are even minor differences, hammer out the expectations that information being shared must be absolutely candid. Hold follow-up audits until satisfied that all information being provided is accurate.

- At this point, sellers must provide a schedule of critical points that either must be held or will be covered up if not observed in a timely manner. All inspectors or other individuals with oversight responsibility have found themselves in a position that one or more key points were missed. Generally, those individuals are placed in a position they must either accept a condition on blind faith or they must demand that work be undone and repeated. In most cases, the individuals are made to feel wasteful for asking that checkpoints be uncovered or that they are a heel for not trusting their "friends." In any case, it is best to understand the schedule and to avoid situations where individuals are asked to ignore their responsibilities.

♦ Once the product is ready for verification and validation, it is best to agree to testing methodology, results, and documentation. With agreement about these elements, it is possible for owners to monitor results based on reviewing test documentation. If documentation shows unacceptable results, the path forward will require rework of earlier steps. If the test documentation portrays only positive results, it would be best to audit some of the results and to actually participate in some of the testing first hand.

Reading the activities just described, you may be thinking, "This sounds like a lot of work. I thought this was going to be an approach that required limited resources." My only response is that nothing is free. It is possible to make some progress toward the desired performance with fewer resources, but it is not possible to accomplish much with no resources.

There is an old story about a farmer who was teaching his cow to eat half as much each day as he did on the previous day. As the farmer concluded, "Things were going well until I taught him to eat nothing. Then he died.

Conclusion

Application of DFR the first time may be somewhat painful because of the steep learning curve. Yet for most members of the engineering staff, the process will soon become second nature. In addition, the results of adding DFR to the engineering development process will soon become such an integrated element of design that folks will wonder how they ever got along without it.

Chapter 8

Obtaining the Cooperation of the Seller

*Promises are the uniquely human way of ordering
the future, making it predictable and reliable
to the extent that this is humanly possible.*

Hannah Arendt

Chapter 7 focused on a situation in which the project team acts as an extension of the owner's organization. In that situation, the sellers' organization is required or compelled to fulfill the owners' objectives when designing a product for future owners. In that situation, the persons designing the asset may require some convincing to apply DFR methodology during the design process. After all is said and done, however, if the individuals performing the design fail to take the appropriate action, the resulting poor performance will reflect back on them in some negative way. Depending on how close or continuous the relationships between the owners and sellers are, individuals who develop products that fail to meet the owners' requirements may be disciplined or discharged.

The same dynamics are not present when owners are purchasing a complete product from the sellers. In this case, the sellers' agents may say their objectives are to meet the owners' (their customers') needs, but that may or may not be true. I have never seen a sales representative or product manager fired or in any way disciplined for producing or selling a product that did not meet the customer's requirements or had poor reliability. In most cases, the issues leading to poor reliability typically led to lower costs and greater profit so the individuals who made the choices leading to the savings were typically rewarded.

Creating a Reliability Partnership

The question to be answered by this chapter, then, is, "How do I convince the sellers of a product to design it in a manner that it will provide the Inherent Reliability, Availability, and Maintainability I need?"

The simple answer is to simply refuse to do business with suppliers whose products do not meet requirements. Although this approach is one alternative, it is not always possible. It is often the best approach because, ultimately, the loss of business will convince the sellers that improvements are mandatory. It is not always possible because occasionally there is only one supplier in a small market. Or if there are more suppliers, it would not be wise to eliminate competition by choosing not to do business with one of the few alternatives.

Therefore, we need to convince the sellers that applying

DFR—and, specifically, DFR that meets the owners' requirements (rather than just the sellers')—will be good for both the
sellers and the owners. The ultimate objective is to create a
partnership between sellers and owners that result in products
that achieve both of their objectives.

When attempting to form this partnership, there are always
altruistic approaches like:

♦ The customer is always right.
♦ Making the customer successful will make the seller
  successful.
♦ A reliable product is good for the economy, the
  environment, the market place, etc.
♦ The first requirement of a quality product is to meet
  requirements.

All of these arguments provide compelling reasons to
include DFR aimed at the owners' requirements in the design
process. They are all true to one extent or another. Yet, many
sellers will not find them to be an adequate reason to comply.
So, what constitutes "an offer that they cannot refuse?"

The best and most simple approach is to build the requirements into the specifications. When sellers choose to bid on
your specifications, they are agreeing to provide a product that
meets the requirements included in the specifications. Make
certain the requirements clearly describe a DFR analysis that is
based on the owners' requirements. Furthermore, the requirements described in the specifications should make it mandatory that sellers deliver documentation describing the results of
the required DFR analysis before delivery of the product. Then,
it will be impossible to meet the specifications without per-

forming the analysis and delivering the documents. If required documentation is delivered prior to acceptance of the product by the owners, the owners have an opportunity to refuse acceptance of a product that does not meet their needs. Although a refusal to accept delivery may not be realistic, owners can withhold some portion of the payment until shortcomings are corrected.

The main challenge is to find the right wording to describe the analysis you wish to have accomplished and the documents you wish to have provided to you before acceptance of the product. Another challenge is to stand your ground when the sellers take exception to the wording that establishes these requirements. A further challenge is to include remedies (for failing to meet the requirements) that are sufficiently distasteful that the sellers will want to avoid them. If the requirements can be ignored without fear of reprisal, it is likely they will be. During contract negotiations, owners should take the position that the DFR requirements are simply a standard part of any comprehensive design process and should be a part of the normal costs. After all, the sellers have been making verbal guarantees for years that their products will be reliable and easy to maintain, with a long, useful life. You are only trying to put those guarantees in writing and document the required performance in clear and unambiguous terms.

Keep Your Objectives in Sight Until They are Achieved

In a recent situation, I spent a great deal of time and effort building the case for using DFR during the design of an asset and doing it in a manner that met the owners' requirements. The sellers expressed a significant amount of enthusiasm concerning the subject. They even created a presentation, a sales pitch, and sales documents describing their DFR process. In effect, the seller said, "We understand what you want and are ready to fulfill your needs".

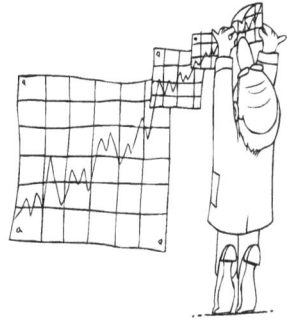

As the time for finalizing the contract approached, the sellers were asked to describe the DFR they would use. The DFR process the sellers presented was one that ensured their own requirements were met (not the owners'). Furthermore, they had significant problems sharing any information concerning components. Although most components are manufactured by third-tier suppliers, they are supplied as replacement parts through the OEM (seller). The sellers add a significant mark-up when they sell the replacement parts. Therefore, they take great pains to prevent the owners from developing a direct link to the parts manufacturers. The sellers are also careful not to allow the owners to have detailed reliability information about the components. The suspicion is that the calculated performance for the combination of parts being provided in the sellers' product will not yield the overall reliability being guaranteed for the asset. If the owners are provided the data, they will be able to perform the analysis for themselves. But without the data, who knows?

Once again we return to having the right wording in specifications to describe the analysis that must be completed to meet the owners' requirements—but there is more. The appropriate wording in the specifications forms the basis for obtaining the desired results. But, it is always possible for the sellers to feign confusion. As a result, there is a benefit in spending some time with the sellers going through the specifications to the point that confusion can no longer exist. As part of this discussion, have the sellers repeat back the meaning of the specifications, providing examples of the analysis they will perform and the reports they will supply.

Eyes On-Sight

Another important element needed to ensure the desired results is to embed a reliability engineer employed by the owners at the sellers' facility during the design and manufacturing process. This may seem extreme in some situations. However, think about the case in which several hundred copies of a single model are being purchased. Suppose the owner is a railroad that plans to purchase several hundred of a specific model of locomotive. It would make sense to provide an on-site resource to overview the design and manufacturing process.

The embedded engineer needs to have the following characteristics:

♦ Knowledgeable about the asset being designed
♦ Familiar with how the asset will be used and maintained
♦ Ability to juggle a number of issues simultaneously
♦ Aggressive and demanding
♦ The kind of person who cannot be frightened or flimflamme
♦ Comfortable dealing with senior executives and escalating contentious issues

By now, you are probably getting the idea that there isn't a great deal of trust in the design process when is comes to the completion of DFR and integrating DFR findings in the final product! Experience has shown two things:

♦ First, many sellers say they go to great pains to ensure their products are reliable, but many do not. Of the ones who do, many proceed in a somewhat amateurish manner. Of the ones who continue in a professional way, many proceed in a manner that meets their own objectives and not those of their customers.

♦ Second, the business model of many sellers is dependent on selling parts and service for the entire life of the products they sell. Highly reliable, available, and maintainable assets do not require either the number of replacement parts or the amount of service of devices with poor reliability, availability, and maintainability. Convincing sellers to do something that will ultimately close a revenue stream is difficult. It is not enough to obtain a commitment; you also need to monitor progress.

The Long Path to Success

Depending on your industry, implementation of DFR in a form that ensures the owners' requirements are met may be an undertaking that will require several years to complete. It is likely to be a long and tedious effort that takes the following path:

1. You try to get it started, but experience pushback.
2. You try again and receive some support, but mostly talk.
3. You try again and receive good support within your own

company, but little acceptance from the sellers.

4. You try again and receive what appears to be support from the sellers, but they find ways around your specifications.

5. Then you try again and have plugged all the holes and finally have a process that appears to be working—but the product doesn't turn out to perform as expected.

6. At last, everyone is working together and the product comes out just like you wanted back at the start.

7. Then you retire.

Chapter 9

After the Purchase is Completed

One of the saddest lessons of history is this:
If we've been bamboozled long enough,
we tend to reject any evidence of the bamboozle.
The bamboozle has captured us.
Once you give a charlatan power over you,
you almost never get it back.

Carl Sagan

The movie *Funny Lady* has a scene in which the male lead, James Caan, is using a stop watch to time a man who is in his office painting a wooden chair with a brush. The chair is sitting on a table in the middle of the office where every aspect of the task is clearly apparent. Although busy talking on the phone, he carefully monitors the person performing the work to see that he is not wasting time during this exercise. Finally when the man finishes painting the chair, Caan stops his watch and goes through some calculations. He finally tells the man that it took him X minutes to paint the chair. He has Y chairs he needs painted and he is willing to pay him Z amount of money per hour for painting the chairs. In total, he will pay the man X times Y times Z for painting the chairs, and not one cent more. The man agrees to paint the chairs for the agreed amount and leaves. Once outside the office, the man who painted the chair yells to someone off in the distance, "Spray 'em."

Changing Results Requires a Different Approach

All too often we spend too much of our time and energy procuring assets the same way the leading man did in the example above. We invest all our efforts in arriving at a fair deal. Then we ignore taking appropriate steps to see that we got what we paid for. It would be very easy to take all the steps described in the first eight chapters of this book and then move onto the next project without ever determining if the asset we obtained achieved our objectives. That would be a mistake—for several reasons. First, much of the reliability of the asset you purchased will come from finding and changing those components and design elements that do not live up to expectations. Second, if you do not monitor performance and hold the suppliers' feet to the fire, the next time you purchase an asset from them, they will know that you don't really mean business; they can get away with taking short cuts.

Even if you have not applied all the steps described in the first eight chapters, there are still steps you can take to see that the asset you purchased meets some acceptable level of reliability performance. Although obtaining the performance you desire may be more difficult if you have failed to take all the right steps leading up to this point, there are generally-held expectations for performance that can be used to enforce your unwritten expectations.

Assuming that you do not wish to suffer the indignities of being duped, the question remains, how can you best track asset performance in a way that allows you to achieve either what has been specified or what you believe you have a right to expect? There is a classic saying that, "All politics is local." There should be an equally accurate saying that, "All reliability is determined at the component level." In other words, to achieve your objectives, you need a system of monitoring reliability that goes down to the component level.

You must begin by monitoring performance at the overall asset level. If sellers have agreed to provide a plant that will produce 30,000,000 gallons of a product per year, and your plant actually produces that amount, then you have little to complain about on the surface. However, if your plant only produces 20,000,000 gallons per year, there is no argument you have a valid complaint.

Suppose your plant produces the required 30,000,000 gallons per year, but does so only after the investment of monumental maintenance and engineering efforts on your part—do you have a valid basis for complaint? You do if you can focus on the specific elements of the plant that are providing unreasonably poor or costly service, then show that they either do not meet the design requirements or reasonable expectations.

The design requirements were established through the design processes described in Chapters 1–8. Reasonable expectations are based on common knowledge about the product or past experience. Suppose you purchase the third in a series of plants that each produces 30,000 barrels per day of gasoline. The first three plants are provided with an electrical distribution system that supports near trouble-free performance. Let's assume your purchase is based on a fixed price contract that allows the sellers to keep the difference between the contract price and their total costs. As a result, the sellers take some shortcuts in the electrical distribution system, causing it to be less reliable than the first two facilities. When you purchased the third plant, your reasonable expectations were that it would be much the same as the first two.

The real question is this: How do you track performance after purchase to be able to get what you believe you paid for? Tracking at the highest levels (e.g., total production and total costs) is a useful starting point, but it is only a starting point. Individuals who have had experience trying to prove the inadequacy of a product using only general measures of performance will tell you it is impossible. When the measures are general in nature, the cause can also be very general. In other words, it is impossible to tell if poor reliability is the result of the sellers' inadequate design or the owners' poor operation.

In my experiences, the only way to quickly drive resolution and produce timely improvements is to maintain a thorough and accurate tracking system. This system must closely track performance in a way that links specific failure modes with the costs and production losses they caused. The reliability engineer must then match the costs and losses with characteristics that were either specified or guaranteed during the DFR process.

In an earlier book, *Failure Mapping*, I described an organized system for defining and tracking the steps leading to failures. This approach allows us to learn from the experience so we can address the specific defects leading to the failures. As part of Failure Mapping, the Malfunction Report is linked directly to the Failure Mode that caused the event. Using this approach, we are able to link specific defects to the costs they cause.

In such a system, the Failure Mode is defined as the Component that failed and its Condition. The practice of recording the Failure Mode in this manner provides owners with a number of capabilities they would not otherwise have.

If the Failure Mode for each event is recorded, the owners will know how much money (in terms of repair costs and lost production) each Failure Mode is costing them. As a part of DFR during the design process, the sellers should have identified:

♦ The failure rate and reliability for the entire asset.

♦ The failure rate and reliability for each component and how those quantities build into the overall asset reliability.

♦ The failure rate and reliability based on each Failure Mode for each component, showing how the total of all Failure Modes builds into the combined failure rate for each component.

If information is gathered as described before, the owners have the ability to quantify the past cost of failures, and then compare those costs to the amount specified or expected. In addition, if information describing both the Failure Mode for each component and the timing of each failure is collected, it will be possible to perform failure growth analysis and forecast how an asset will perform in the future. Future performance expectations that are based on established analysis methods provides owners with the ability to forecast poor performance and take steps before the unacceptable performance occurs.

The most recognized form of statistical failure analysis is Weibull analysis. One element of Weibull analysis is the percent of the total population of a specific component that has failed at any point in time. For instance, if you operate a fleet of vehicles that contain a total of 50 of a specific device, every time one of those devices fails, the percent of the total population that has faied will increase by 2% or more, or one-fiftieth of the population. A second element is the life at which each of the failures occur.

These two measures are then plotted against each other on logarithmic scales. Weibull analysis is useful in forecasting the cumulative number of failures that should be expected at any time. It is also useful in determining the usable life of a component. The example below describe several simplified mock-ups of Weibull charts and the information that is available from them.

An inherent weakness of Weibull analysis is its inability to discern the failure rate resulting from various Failure Mechanisms. That is the case, unless the failures resulting from various Failure Modes are plotted individually.

In other words, if all the different kinds of failures of a component are plotted together, the results of the analysis will point to a cumulative failure rate. If the cumulative failure rate is the result of several causes, then addressing only one of the causes will result in only a partial improvement. If the different failure modes are tracked separately, it will be possible to determine the portion of improvement that will result from addressing each cause.

The following figures provide simplified mock-ups of Weibull charts. Figure 9.1 represents the results the results that would be portrayed if all Failure Modes are combined.

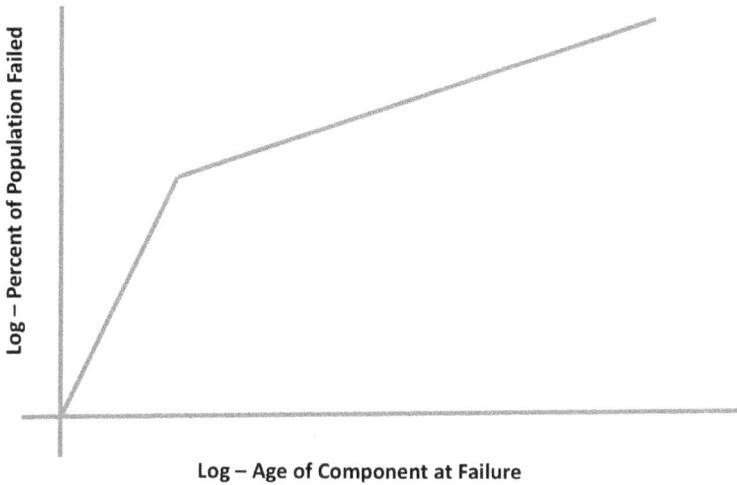

Figure 9.1 Combining all Failure Modes

Figure 9.2 shows the result of one portion of the failures in Figure 9.1. For example, it might represent the random but relatively constant failures due to wear that would be experienced over the entire life of the component.

Figure 9.3 represents the other portion of failures. In this case, the steep initial slope represents some form of infant mortality. Once the early life failures are done, the relatively flat region represents a long period of relatively few failures.

The value of distinguishing failure modes within the Weibull graphs is the certant you are making the right repairs at the right times.

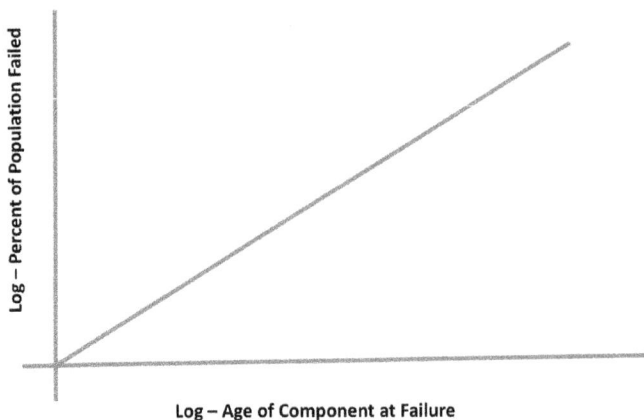

Figure 9.2 Showing Random Failures over the Asset Life

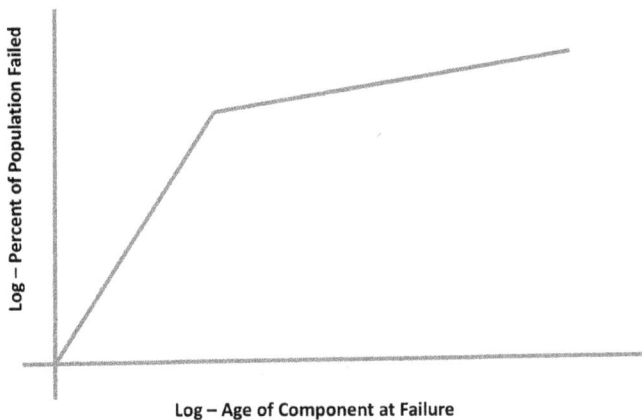

Figure 9.3 Infant mortality followed by relatively few failures

Suppose a wire is in a service where it is occasionally overloaded, leading to failure. Assume the same wire has been installed in a manner that exposes it to abrasion, causing wear and ultimately failure. If a different larger wire were installed, the portion of failures resulting from overload would be eliminated. If the new, larger wire was installed in the same manner as the old wire, it would still be subject to abrasion, wear, and failure. If repair work orders were closed by recording the parts identification of the wire only, the analyst or troubleshooter would be left to determine how best to address the problem.

If both the component name and the component condition were recorded, it would be possible to identify both causes. For instance, Current Supply Wire—Overheated would lead the analyst to look for an overload condition. Also, Current Supply Wire—Abraded would lead the analyst to look for signs of poor installation or rubbing.

The statistical frequency of each Failure Mode would lead analysts to look for two solutions if they wanted to completely eliminate the problem.

For example, assume that over a ten-year period a specific wire has experienced the failures described in Figure 9.4.

Ten-Year Wire Failure History		
Year	Overheating Failures	Abrasion Failures	Cumulative Total
1	2	5	7
2	3	6	16
3	4	7	27
4	2	5	34
5	1	4	39
6	4	6	49
7	5	5	59
8	3	8	70
9	2	3	75
10	1	5	81

Figure 9.4 Ten-year wire failure history

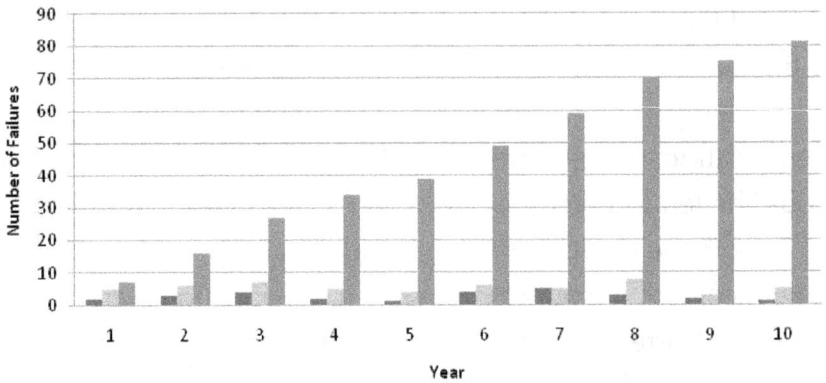

Figure 9.5 Cumulative wire failures

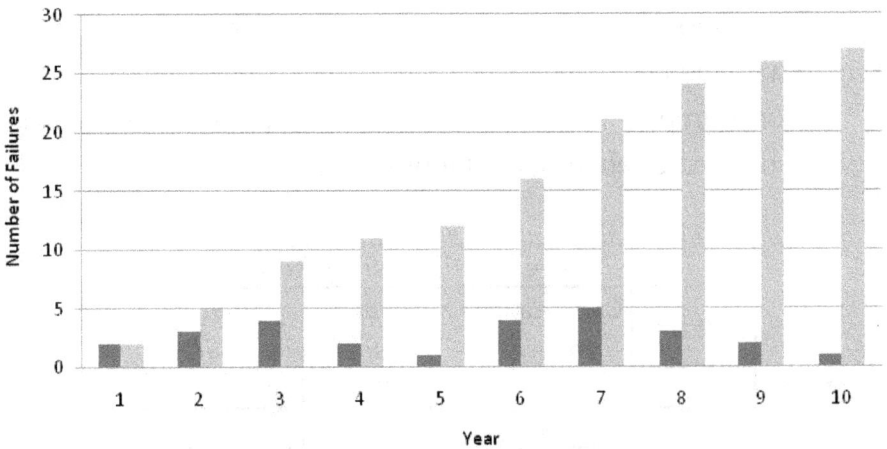

Figure 9.6 Overheating failures

Graphing the data from the table would result in the bar chart seen in Figure 9.5.

If you looked only at the cumulative number of failures, you would find an unacceptable number of failures, but you would not know the source of failures. However, if you sepa-

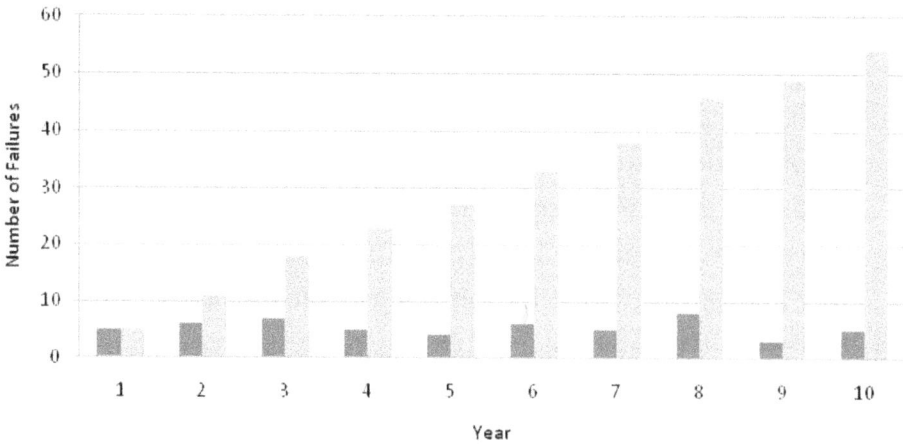

Figure 9.7 Abrasion failures

rated failures into the two distinct sources, as shown in Figures 9.6 and 9.7, you would have a much clearer idea of the source of problems and where you should invest your efforts to eliminate the failures.

Separating failure modes would show that the majority of failures are resulting from abrasion or wear. A simple solution might be simply to provide some protection or reroute the wire.

Linking Failure Mapping to DFR

Now, let's see how Failure Mapping links back to the manner in which DFR is accomplished.

In this example, it is expected that neither the loss of macro-reliability of the overall system nor the specific loss of micro-reliability of the wire would be reflected in the reliability calculations made by the sellers or the information provided by the sellers to the owners. Designing a wire for a smaller load than is expected would be a design mistake. Also installing a wire in a manner that would subject it to abrasion would be a manufacturing error. Failures of either kind should "stick out like a sore thumb" when the owners compare actual performance of the asset to the kinds of failures described in the reliability analysis provided by the sellers. Because these kinds of failures are typically not found in the analysis provided by sellers, their impact is highly visible. Clearly, overheating of a wire would signal a failure of one of the most basic engineering steps. Also abrasion and wear of a wire would also signal a failure in the manufacturing process both during the assembly step and during the quality control step.

Neither of these Failure Modes were included in the DFR analysis (in terms of a reliability debit or owner cost). Therefore, the owners would have a right to expect the sellers to correct the problem at no cost.

On the other hand, some components are expected to have a statistically small but still existent number of failures. An example may be a pressure sensor. Modern pressure sensors work as the result of a small microprocessor mounted on a flexing diaphragm. As the diaphragm flexes, the properties and signal sent by the microprocessor changes. Because the diaphragm continually flexes, it does not have an infinite life. At some point, it will fail due to fatigue. Fatigue failures are a kind of failure that is spread across a normal distribution, with a few occurring early, the majority of the population occurring toward the center of the distribution, and a few occurring late.

It would be reasonable to expect that some small number of pressure sensor failures will occur early during the life of a population of such devices. On the other hand, if more failures begin to occur than is statistically reasonable, it is important to understand if the expected life of the population is less than expected or guaranteed. If so, the weakness should be identified and the sensors exchanged with ones having an acceptable life.

Pressure sensors have a finite life that is associated with the fatigue life of the internal diaphragm. However, they may also be caused to fail for other reasons. For example, if the annulus that contains the diaphragm becomes plugged with debris, the sensor might fail due to plugging or even concentration cell corrosion. If the stream where the sensor is located is either more acidic or basic than expected, the diaphragm may experience cracking if it is made from an inappropriate material.

Again, the condition of the failed component must be recorded with the component's description when the repair job is closed. Otherwise, it will be impossible to determine if the Failure Mode is one of the types that were expected as a result of the DFR analysis. If the number of failures resulting from an expected failure mode is greater than the expected number or if the Failure Mode is a type that was never considered during the design, the owners need to know this fact. They can then either seek corrective action from the seller or add some form of PM/PdM to their proactive maintenance regimen to intervene before failures can occur.

In the process of linking Failure Mapping to the DFR for the Owners, it will be the responsibility of the owners to take the following steps:

1. Obtain the macro-reliability from the sellers.

2. Obtain from the sellers the micro-reliability for all components included in the calculation of the macro-reliability. The seller should be able to tell the owners the specific Failure Modes that are included in each component's micro-reliability.
3. Closely track the Failure Mode using Component and Condition when closing all failures to the asset.
4. Calculate both the overall micro-reliability for each component and the micro-reliability associated with for each Failure Mode for each component. (This will be helpful in identifying Failure Modes that the sellers did not take into account when performing DFR.)
5. Regularly compare the results of analysis completed in Step 4 above with the information provided by the sellers in Step 2. Report any inconsistencies back to the sellers.
6. Depending on the significance and severity of the disparity, ask the sellers to take corrective action.
7. If the sellers do not take corrective action, the owners must provide some form of PM/PdM to address the new or earlier failure mode.

Few owners have an engineering staff equal to the organizations that design and sell complex assets. It should be within the capacity of the owner to identify situations when the asset and component performance does not meet expectations and requirements. Still, it is often outside the owner's capacity to re-engineer the components. Linking expectations to performance is a direct method of both achieving desired reliability and forcing the sellers to provide the owners with what they have paid for.

One valid question is how best to achieve desired performance in situations when the owners did not do everything right throughout the design process. In other words, if the designers did not apply DFR during the design, how does they make things right?

Despite the fact that DFR was not used to create a design that met all the owners' requirements, and despite the fact that the data resulting from seller-performed DFR is not available to the owners, it is possible to gather data and perform an analysis to determine the capabilities of the system the owners received. In most cases, owners are provided with data books containing detailed descriptions of each key component or equipment item. Increasingly, the data sheets contain the results of reliability testing. They also provide both designers and owners with the expected reliability and usable life of the components. If the data books do not contain that information, most suppliers have the data and will provide it when asked.

Although it is not a simple matter, it is possible to either construct a simplified Reliability Block Diagram or a simple reliability allocation scheme (sum of all individual failure rates leading to a composite failure rate) using the individual component reliability information described above. Doing so will provide owners with a rough idea of the reliability they should expect.

This information can be used in two ways. First, if the actual performance is less than expected, it can spur the owner to identify specifically which component or components are underperforming. Once it is known which is underperforming, it is possible to determine if the component is:

1. Failing in the same manner as expected by the suppliers, but at a higher rate.
2. Failing in an unexpected manner, one not used in the suppliers' failure analysis.

In either case, the owner can contact the component supplier to find a solution that will lead to acceptable performance.

The second way the information coming from the RBD or reliability allocation scheme can be used is to determine if the device will achieve the owners' requirements as currently constructed. The number coming from these calculations is something close to the inherent reliability. If properly operated and properly maintained, the device will achieve a reliability performance close to the calculated number. It will not exceed it by much. If the calculated reliability is not adequate, the owner can use the analysis to identify the components that must be either supplied with redundancy or replaced with more robust components to perform adequately.

Once the modification has been made, it is necessary to use the information coming from the Failure Mapping system to track actual performance of the individual components. If the revised system does not perform up to expectations, it will be necessary to identify which of the current components are not performing up to expectations either because of anticipated Failure Modes occurring at a higher than expected rate or unexpected Failure Modes. In either case, the owner will need to return to either the system sellers or the component suppliers to have the situation rectified.

Another valid question is how do members of very small organizations with limited resources make use of these recommendations to improve reliability? This is one of those questions like asking how it looks on the other side of the mountain. You won't know and may not believe me until you get there.

First, on the whole, it requires fewer resources to be proactive than it does to be reactive. It requires fewer resources to specify and purchase reliable assets than it does to deal with unreliable assets after they are delivered.

Second, concerning the time and effort it requires to establish and to maintain the information in a Failure Mapping based system—again, once it is established, it takes less time than your current systems.

Specifically:

♦ Once you have identified the possible Malfunction Reports in terms of Impaired Function and Specific Behavior, it takes less time and effort to report a failure than to write a paragraph describing the failure.

♦ Once you have identified the small number of Failure Modes that are associated with each Malfunction Report, it is far easier for the person performing the repair to select the appropriate Failure Mode from a list than it does to come up with the words needed to describe the repair.

♦ Ultimately, the link between each Malfunction Report and the possible Failure Modes provides information critical to accurate diagnosis, triage, and troubleshooting. The data helps make your maintenance department more effective and efficient.

♦ The data collected using Failure Mapping provides all the information needed to perform accurate FMEAs, identify specific component failure rates, perform component failure growth analysis, and identify Failure Modes and Failure Mechanisms.

When resources are most scarce, the effective use of information to leverage their numbers is most important. I clearly understand that initial implementation of these initiatives requires more resources and those additional resources create a "hump" in the way of progress. On the other hand, the view on the other side of the hump is pretty sweet.

Conclusion

It so happens that I am writing this book during football season. Being a football fan, I watch a lot of games and hear a lot of the coaches being interviewed after their teams lose. It seems one of the most common things that losing coaches say is that their teams "failed to finish." By that they mean that their teams practiced well and played well through most of the game but simply did not play all the way to the end of the game. This chapter is intended to emphasize the things needed to properly "finish" the activity of procuring a new asset.

It is impossible to tell for sure that sellers have met their commitments until you have operated the asset for some period of time. By that time the sellers are likely to have "folded their tents" and moved on—it will be difficult to gain their attention once again if there is a problem. Despite the difficulty, it is incumbent upon owners to do whatever is necessary to obtain the product they believe they paid for.

Chapter 10

Conclusion

The self is not something ready-made, but something in continuous formation through choice of action.

John Dewey

At the beginning of any venture that entails the investment of large amounts of capital into a complex asset, owners have a certain vision in their mind. Clearly, they would not enter into the venture and all the associated risk if this vision seemed likely to end in losses and an uncertain future. The vision must include an asset that:

♦ Costs an amount very similar to the amount originally estimated
♦ Is completed very close to the time described in the schedule
♦ Produces the anticipated product at the expected rate
♦ Costs very close to the anticipated amount to operate
♦ Costs very close to the anticipated amount to maintain
♦ Is reliable, available, and maintainable

Questioning the Ability to Produce a Reliable Design

Recently, I participated in a meeting discussing the need for a major supplier to enhance the reliability of his product by adding focus to his Design For Reliability process. In that meeting, a senior executive of the company who had a technical background asked if it was possible for the seller to perform the analysis and describe the expected reliability (availability and maintainability) performance over the entire life of the asset. Although the seller tended to dance around and not really answer the question, I found two elements of the issue quite interesting.

First, the individual did not ask the same kind of question concerning other elements of the asset design. He did not ask if the seller could perform the analysis that would perform the needed functions over its entire life or if the seller could produce an asset that could operate safely over its entire life. Nor did he ask if the asset could be operated and maintained at an acceptable cost over its entire life. He only asked if the seller could perform the analysis needed to ensure the various reliability related characteristics for its life.

I found this point intriguing because it highlights an assumption that one's ability to perform reliability analysis is much less capable than design for other key asset characteristics. There is an enigma in that assumption. If sellers are unable to perform a design that assures reliability, they must also be unable to perform a design that assures all those other characteristics.

Questioning the Ability to Sustain Performance over the Entire Life

Second, there is another question hidden within the responses expected of the first question. That question is: "How accurately can you describe the reliability, availability, and maintainability performance as time passes over the thirty year life of the asset?"

This question is more intriguing than the first because it highlights a number of other issues:

1. How accurate are these calculations? In many ways, the analysis of reliability, availability, and maintainability are like detailed budgets and project schedules. In the case of both budgets and project schedules, it is clear that one is attempting to provide a clear picture of how he expects the future to look. In addition, both are providing a guide that highlights a direction for future activities. (For instance, you do not spend money on unbudgeted items or perform work that is not on the schedule.) But most important, recognizing that things change in real life, these tools for describing the future provide a starting point from which real life events can depart and one can measure the difference. Lacking a starting point, it is impossible to tell how far one is off course and why. The same is true of the products coming from reliability, availability, and maintainability analysis. They provide a starting point for understanding how the asset should perform and a tool to look for differences if and when it does not. It provides a basis for improvement.

2. How will changes over time affect the accuracy? Clearly if all the assumptions made during the DFR process remain unchanged over the life of the asset and the asset does not

perform as advertised, the owner will have the right to complain. On the other hand, if the conditions in which the asset is used change dramatically, the owner will have less of a basis for complaining.

3. How will the sellers and the owners use the information included in the forecasts? Many of us have worked with individuals and companies that always take the position the other person is always wrong. If you did exactly as I asked you to do and things went wrong, you should have been smart enough to know that I gave you bad instructions. You are the expert; you should have taken steps to keep me out of trouble, so you are to blame. We also know people who will find ways to reinterpret any words or any contract to their advantage. If the sellers do not trust the owners to use the DFR products in a mature and ethical manner, it is likely they will withhold information or provide it only in a manner that protects the sellers from future liability. This situation is likely to minimize the value of the information being shared.

4. How willing are you to invest scarce resources on issues that will not come into play for more than twenty years? During economic downturns, many companies get into a position where they do not spend resources unless they have a one- or two-year payout. In other words, unless the Return on Investment is large enough to provide pay back in the immediately visible future, the investment is not made. Unfortunately, repeated economic downturns have taught some companies to operate in that manner, even when times are good. This habit results in companies buying assets that are capable of performing reliably for only a short period of

time. Although the overall asset may be expected to be a part of the thirty-year business plan, many of the decisions made on key features of the asset will be based on much more short-sighted analysis.

As for the executives' question, "Is it possible to perform the analysis that will describe the reliability, available, and maintainability for the thirty-year life of an asset?? the answer is "Yes." But that response comes with several caveats, including the following:

♦ All the participants need to play like mature adults. By this I mean that representatives of both the sellers and the owners must take into consideration the fact that no form of analysis is perfect. It is possible that some aspects of the analysis may be flawed; the reason is neither incompetency nor unethical behavior. Although the businesses of the sellers and the owners are different, it is in the best interest of both to solve the problems and create an asset that is reliable, available, and maintainable. Then both need to act in a transparent manner and share all the available information.

♦ Everyone must recognize that the forecast has a number of inherent assumptions. Some may not have been clearly written down. However, that fact does not preclude the fact that they exist and were a part of the design premises. Again, if poor assumptions existed at any point in the design and they led to problems, the best approach is complete candor.

♦ Key elements of the design and manufacturing process are likely to change. Even with the best of intentions, someone in the supply chain or the manufacturing process will make some change that will produce a negative result. When this kind of change occurs, it is best to identify the change as quickly as possible and to do whatever is needed to make it right.

♦ When elements that affect the design or manufacturing process but are beyond our control change, we will be best off if we adapt. Progress is a constant. Sometimes progress produces changes we would rather not see happen because they make our lives more complicated. Although that is unfortunate, it is a part of life. Your best response is to develop resiliency to change. Adapting does not mean diluting your requirements. Instead, it means meeting your requirements in a new way.

♦ The owners and sellers are in different businesses and have different business models. Despite the fact they make money in a different way, both the sellers and the owners can be successful if things will work out best they choose to act like partners and support the others business needs.

Reliability is All about the Owners and Their Assets

As a final point, it is important to realize that some rather large companies out there have decided that it is all about them. These companies seldom find the midpoint in any form of cooperation. It always has to be their way.

If you have no choice but to deal with this kind of a major asset supplier, it is important to understand that you, as the long-term owner, still need to demand your requirements are met. The relationship may not be as cordial as the one described above, but you will live through it. In the end, you may have a few less hairs and the ones you retain may be a little more gray, but you will survive.

Appendices

1. Detailed Description of Life Cycle Cost Elements
2. Typical Owner Specification for DFR
3. Example of Reliability Analysis Report
4. Example of Availability Analysis Report
5. Example of Maintainability Analysis Report

Appendix 1
Detailed Description of Life Cycle Cost Elements

This appendix describes several components of life cycle costs. These descriptions should help the reader understand elements that should be included in the various cost categories.

Initial Cost

For sellers, the initial cost of an asset is their opportunity for success. Pricing for their products should include a valuation for all the "good will" they have built over the years through advertising and reputation. To the sellers, the sales price should reflect the price being charged by competition and the relative placement in the market enjoyed by each competitor.

In some cases, issues totally separate from asset value will play a role in determining the price. For instance, if a company is viewed as having a poor safety or environmental record might require a negative "good will" adjustment to pricing. In effect, they must use the value of their products to buy back the "good will" they have lost. It is incumbent upon the owners to understand this issue if they are to benefit from it.

For the buyers, the asset being purchased is simply another burden. The new asset provides an increased opportunity for added production, but also introduces added risk of loss. The investment in a costly asset is the final result of comparisons with other investment opportunities. It includes an implicit assumption that the asset will perform as advertised, in terms of both its ability to produce and its attendant costs. If the asset fails to perform as expected, it would have been better to take the money and put it in a bank account bearing some minimal interest. The value associated with the sellers' "good will," past advertising, and reputation is of little or no importance when an asset is costing more or producing less than the amounts used to justify the investment.

First costs frequently seem far more tangible to the individuals making purchasing decisions than the costs that come later in the life cycle of an asset. It is important to introduce the concept or LCC and Total Cost of Ownership (TCO) long before the decision process on a new asset becomes necessary. Decision makers need to be provided with clear examples of how shortsighted decisions on smaller assets have introduced unnecessary costs.

Consider a supply organization that chose to purchase high pressure fuel lines for a diesel engine from an unrecognized supplier. Based on the number or these fuel lines being replaced every year, the anticipated savings were expected to be several thousands of dollars every year. It turned out that the replacement fuel lines did not fit correctly, leading to fuel leaks and ultimately engine fires. The cost of the engine fires totaled several million dollars before the defective fuel lines could be found and removed from service. It was impossible that the savings from this change could ever cover the costs of losses.

When meetings were held with the purchasing organization to discuss this issue, the senior supply representative said that their organization needed to continue looking for opportunities like this one to reduce the Total Cost of Ownership. That comment served as an introduction to a long and fruitful discussion concerning the real meaning of TCO.

Operating Costs

All assets have some operating costs. Generally speaking, the costs of operating expenses are part of what is assumed when calculating the profitability of an asset. Although sellers are typically not accountable for paying operating costs, they may find themselves being held accountable if the operating costs are different than was advertised or guaranteed. Assume that poor reliability of a plant caused the owner to need to provide extra operators because of frequent shutdowns and start-ups. The owner would have a basis for a complaint and either compensation or corrective action by the seller. Assume that you purchase a fleet of vehicles and their fuel and oil consumption are far greater than advertised. Again, the owner would have a basis for a complaint and the seller could be held accountable. Both of the examples would depend on the requirements set during the early phased of DFR and a documented agreement between seller and owner.

One might say that an element like the cost of fuel or raw materials are not strictly a part of DFR and should already have taken into account during conventional design steps. That might be true. However, conventional design analysis frequently misses debits that are the result of poor reliability of the asset. For instance, if the sensors that play a critical role in the efficiency of an engine tend to drift as they age, the engine will have reduced efficiency until they are replaced. The replacement of these sensors is an addition to the TCO and the availability of the asset while the engine is down for replacement of the sensors in question.

Maintenance Costs

Some planes are known as "hangar queens" because they spend a great portion of their lives in a hangar being maintained. In addition to the lost profit opportunity, an unusually large amount of maintenance introduces unexpected costs. Again, there are certain performance characteristics that are openly discussed when negotiating the development of a new asset. Occasionally, owners forget to discuss just how much work will be required on their part to obtain that performance. That omission is one that will be regretted for a long time.

In addition to quantifying the reliability, availability, capacity, and efficiency of the asset, it is important to quantify the amount of work that is necessary to deliver that performance. The amount of work necessary includes required predictive and preventive maintenance as well as anticipated repairs and overhauls. The seller will be accountable for a portion of these costs during the warranty period. If the owner has taken the time to clearly define requirements and the seller agrees, the seller may also retain some significant portion of responsibility for these costs for the long-term.

Occasionally, a significant component (e.g., piston rings, main bearings, or connecting rod bearings) may have a usable life that is significantly less than the overhaul interval. In that case, the overhaul interval may need to be reduced, leading to the seller being accountable for a significant portion of the overhaul cost.

The point if this discussion is to highlight the extent to which owners and sellers should go in determining performance requirements and ultimate responsibilities for ensuring those requirements are met. Silence on these issues during the period of negotiation before a sales agreement is completed works to the advantage or the seller. In these cases, the sellers are responsible only for delivering the asset and little else beyond that point.

Costs Related to Failures

The objective of discussing requirements associated with reliability performance is clarity on the asset's value in terms of what the asset will be able to produce and the costs associated with failures. If we were to analyze the total sales of assets in the world, we would find that most assets are sold with little or no agreement on the expected reliability. We would find a relatively small portion for which requirements associated with reliability are established. We would find an almost infinitesimal portion for which the allowable costs associated with failures is addressed and for which responsibility is agreed. In most cases, once an asset is sold, the owner is stuck with the cost of failures—end of discussion.

Sellers will never become truly interested in reliability until they are held accountable for the cost of failures. Some amount of unreliability and a small number of failures are to be expected, and owners should plan for them in their business model. However, when the quantity of failures exceeds a certain number and the downtime exceeds a certain amount, they go beyond any realistic allowance in a business plan. In that case, either the sellers did not understand the owners' requirements or they did not provide an asset that met those requirements.

It is important for both the owners and the sellers to clearly understand reliability requirements. It should be expected that if these requirements are not met, action will be taken to ensure correction is pursued.

Costs Related to Planned Outages or Overhauls

As discussed above, most capital assets have a time when some form of overhaul or renewal is necessary. The cost of these events in terms of lost availability and maintenance costs make them significant elements of the owner's business model. If they are more expensive or more costly than envisioned, the asset may not deliver the expected return on investment. It is critical that the owner make requirements in this area clear to the seller.

The importance of this requirement is particularly clear in situations when owners are dealing with a one-of-a-kind manufacturing facility where the world production of a commodity stops when the unit is down. Although you may think this would be a truly unusual situation, it happens more often that one might expect. It also seems that the one-of-a-kind plants are typically difficult to operate and hard to keep running. (If they weren't, then everybody would be doing it.) As a result, it is particularly important to ensure that facilities of this kind are built in a robust manner to survive reliably from one overhaul to the next.

Appendix 2
Typical Owner Specification
for DFR

The following provides a typical example of the narrative owners may wish to include in their specifications. Using this kind of narrative should ensure that the form of analysis needed to meet their needs is completed and that the information needed to take full advantage of that analysis is provided in a usable form. (The example provided contains details associated with the purchase of a fleet of locomotives. Those details will need to be modified for other assets.)

DFR—Basis of Analysis, Design Choices, and Component Selection

This document is intended to provide a description of the analysis and products needed to meet the owners' needs concerning Reliability, Availability, and Maintainability of a new product. It is intended to accomplish two objectives that will directly benefit the owners. First, the documentation provides a level of assurance that the sellers have performed the analysis required to ensure the required performance. Second, it provides the owners with a clear description of the steps they are required to take to ensure the advertised performance. The objective is to reconcile the expectations for both the seller and the owner concerning reliability, availability, and maintainability performance and the part each must play.

Requirements

- Expected Reliability
  - Definition of a Failure Loss of a Critical Function
  - Reliability Performance Requirement = 1 Road Failure / Locomotive-Year for the entire fleet
- Expected availability
  - Planned unavailability
    - Describe all proactive tasks that can be identified in advance
    - Include Typical Task Time and Waiting Time
  - Unplanned availability
    - Describe typical task required to repair event resulting from 1 FLY failure
    - Include Typical Task Time, Waiting Time, and Transit Time
- Expected maintainability
  - Restoration of Inherent Reliability
    - Tools, skills, and testing equipment needed to identify weak or failed components
- Ratability
  - Amount of task time needed to complete repair and restore inherent reliability including issues determining repair time.
- Details describing basis of analysis
  - Usage/Mileage
    - MWH/month
    - Mileage/month
  - Service
    - Coal
    - Distributed Power

- Live Renewal/Extension
  - Maximum of two overhauls
  - No renewal-do not assume that the owners will perform life renewal tasks beyond the scope of a typical overhaul in which current systems are restored to like new conditions

Analysis

- The DFR will include Reliability Block Diagram and Reliability Centered maintenance Analysis using Isograph, Reliasoft, Relex, or equivalent software. The selected software must be capable of performing life cycle simulations of reliability and availability performance and calculating overall reliability, availability and life cycle costs. The selected software must be capable of creating reports that clearly portray results for:
- Overall asset reliability
- Component performance (Weibull life) based on PM/PdM/ Replacement assumptions.
- Component PM/PdM/Replacement requirements for entire 30-year life
- Assumptions affecting responsiveness including spareparts availability and workforce responsiveness.
- Utilize ability to iterate back and forth adjusting reliability, availability, and life cycle costs based on varying component selection, system configuration, and maintenance programs at each annual interval of the 30-year life.

Products Delivered Before Starting of Fleet Manufacture

♦ RBD Analysis—Deliver product of complete RBD analysis in electronic file form

♦ RCM Analysis—Deliver product of complete RBD analysis in electronic file form

♦ Overall System Reliability—Deliver report showing overall asset reliability over the entire 30-year life including TCO and LCC comparisons.

♦ Component Reliability—Deliver report showing overall asset reliability over the entire 30-year life including TCO and LCC comparisons for each of the components finally chosen to be included in the delivered product.

♦ Availability—Provide a bar chart covering each year in the 30-year life of the asset. The bars included should describe down time for planned maintenance (including down time for regulatory requirements), unplanned maintenance (failure repairs), and overhauls.

♦ Maintainability—Provide a complete list of all tasks anticipated over the 30-year life. For each task, describe the basis of the requirement (Failure Mode and Failure Mechanism), what is required to identify defective component and ensure functionality is restored, and task time required to perform the repair.

Observation Concerning Cost Effectiveness and Expected Performance

In order to avoid failures, sellers usually take advantage of tools that allow prevention and early identification of deteriorated and failing components. They would rather use these tools than depend solely on selecting the most robust (and expensive) components or on a configuration integrating a significant amount of redundancy to achieve required objectives. This assumption is an important element of controlling first cost while achieving lowest TCO. Seller will likely include the following elements:

- Prediction—This element includes direct identification of conditions that exist in sensing or control systems. Prediction provides advanced notification when deterioration has impacted capabilities but before failure has occurred.

- Detection and Diagnosis—This element includes sensing and reporting that will identify the specific failed component and its condition. This identification allows advanced preparations to be made, triage to be accomplished, repairs to be made in the most effective and efficient manner possible, and engineering analysis to be accomplished in a timely manner that identifies components failing differently or more frequently than expected.

- Troubleshooting—This element includes a system for prioritizing repairs when more than one possible cure exists. Again, the objective is to improve availability and maintainability by eliminating uncertain repair steps.

It is critical that these kinds of elements, which are being proposed as part of the system used to achieve required performance, are both tangible and proven beyond question. If they are not current examples of their successful application, they should not be included. Said another way, if the sellers cannot provide current examples of situations where prediction, detection, diagnostic, and troubleshooting tools are producing enhanced reliability, the value of these tools should not be included in reliability or availability calculations.

Appendix 3
Example of Reliability
Analysis Report

This appendix provides a typical example of the final report covering the Reliability Analysis that owners may wish to request from the sellers of an asset. The report will help ensure the information needed to take full advantage of that analysis is provided in a usable form.

The Reliability Analysis Report should begin with a cover sheet describing the content of the report. The next element of the report is a table showing the calculated reliability for each year of the asset's life. Figure A3.1 provides a rough example.

The next element of the report is a spreadsheet showing the reliability for each component that was used as part of the Reliability Block Diagram calculations. Figure A3.2 provides an example.

The report should include an appendix that has two attachments. The first attachment should be a copy of the complete RBD calculations or electronic file. The second should be a complete set of the Weibull graphs or other analytical forms used in determining the component reliability that was used.

Asset Description - Distilling Unit

Year of Life	Calculated Asset Reliability for the Year	Description of Failure 1	Description of Failure 2	Description of Failure 3	Description of Failure 4
1	95%	Charge Pump Failure - Infantile Bearing Failure - 2.5%	Internal Heat Exchanger Leaks - Improper Gasket - 2%	Electrical Trip - Undersized Fuse - .5%	
2	99%	Reflux Pump Failure - 1%			
3	100%				
4	100%				
5	100%				
6	99%	Reflux Pump Failure - 1%			
7	96%	Heat Exchanger Plugging - 2%	Tray Fouling in Crude Column - 2%		
8	100%				
9	100%				
-					
-					
-					
28					
29					
30					

Figure A3.1 Calculated Reliability Over an Asset's Life

Component Reliability Used in RBD Analysis with Maintenance and Replacement Assumptions										
							Required Maintenance and Replacement			
Asset Name	System/Function Name	Component Name	Annual Reliability	Failure Mode 1	Failure Mode 2	Failure Mode 3	Required Maintenance	Maintenance Interval	Required Replacement	Replacement Interval
Asset XYZ										
	System/Function									
	Braking									
		Brake Shoe	0.99999	Worn	NA	NA	Inspection	Annual	Upon Condition	Expected 4 year life
		Brake Drum	0.999999	Worn	NA	NA	Inspect	Annual	Upon Condition	Expected 12 year life
		etc.								
	Engine									
		Main Bearings	0.9999	Worn	NA	NA	Oil Analysis	Quarterly	During Overhaul	Expect 12 Year Life
		Wrist Pin Bearing	0.999	Worn	NA	NA	Oil Analysis	Quarterly	During Overhaul	Expect 12 Year Life
		Injector	0.99	Regulatory Requirement	Plugged	NA	NA	NA	Regulatory Mandated Cahnge	3 Year
	etc.									

Figure A3.2

Appendix 4
Example of Availability
Analysis Report

This appendix provides a typical example of the Final Report covering the Availability Analysis. Owners may wish to request this report from the sellers of an asset to ensure the information needed to take full advantage of that analysis is provided in a usable form.

The Availability Analysis Report should begin with a cover sheet describing the content of the report. The next element of the report is a spreadsheet showing the calculated Availability for each year of the asset's life. Figure A4.1 provides an example.

Each event shown during the life of the asset should be linked to an event in either the Reliability Centered Maintenance model or the Reliability Block Diagram model. As a result, this section of the report should include an appendix with the following items:

1. A copy of the RCM analysis highlighting each form of Predictive or Preventive Maintenance that will result in a planned outage event.

2. A copy of the Reliability Block Diagram highlighting each event that will result in an unplanned outage.

Annual Availability for Asset XYZ

Year	Planned Outages			Unplanned Outages			Cumulative Downtime for this year
	Planned Event	Run Interval Limiter	Duration of Outage	Unplanned Event	Triggering Failure	Duration of Unplanned Outage	
1	Initial Inspection	6-month Run	24 hours				24 hours
1	First annual Inspection	12-month Run	24 hours				48 hours
1				Calculated Instrument Failure	1 Failure per year from one of 50 instruments	24 hours	72 hours
1				Calculated Pump failure	1 failure per year from one of 20 pumps	48 hours	120 hours
2							
2							
3							
3							
4							
4							
5							
5							
.							
.							
29							
29							
29							
30							
30							
30							
						Average Annual Unavailability	

Figure A4.1 Availability Analysis Report

Appendix 5
Example of Maintainability
Analysis Report

The following provides a typical example of the Final Report Covering the Maintainability Analysis. This is a report that owners may want to request from the sellers of an asset to ensure the information needed to take full advantage of that analysis is provided in a usable form.

The Maintainability Analysis Report should begin with a cover sheet describing the content of the report. The next element of the report is a spreadsheet describing the results of the maintainability analysis for each task that is recognized as being needed over the life of the asset. Figure A5.1 provides an example.

To help the owner better understand the basis for the analysis and how it was accomplished, an appendix including the following elements should be attached:

1. A copy of the electronic file containing the RCM analysis. This appendix should include the key used to link the line items of the analysis to the line items of the above spreadsheet.
2. A description of how various maintainability analyses were completed.
   a. Was the analysis completed using an actual walk-through using a similar model?

Maintainability Analysis Report

System/Function	Component or Equipment Item being Maintained	Proactive or Reactive Task	Results of Task (Same as Old/Same as New)	Basis for Determining the Duration of the Task	Basis for assuring that the Inherent Reliability is Restored by the Task
Braking					
	Brake Shoes	Inspect Brake Shoes	SAO	Observed Current Practice	Standard Method for Measuring Shoe thickness and condition exists.
	Brake Computer	Exercise Internal Diagnostics	SAO	The Duration is set by the brake computer software	Test devices are provided that verify condition or computer boards.
	etc.				
etc,	etc.				

Figure A5.1 Maintainability Analysis Report

   b. Was the analysis completed in an office using drawings
       only?
   c. Was it accomplished in some other manner?

References for Further Reading

1.Crowe, Dana and Feinberg, Alec; Design for Reliability (Electronics Handbook Series); CRC Press, Boca Raton, FL; 2001

2. Bazovsky, Igor; *Reliability Theory and Practice*; Dover Publications Inc., Mineola, NY; 2004

3. Daley, Daniel T.; *The Little Black Book of Reliability Management*; Industrial Press, New York; 2007

4. Daley, Daniel T.; The Little Black Book of Maintenance Excellence; Industrial Press, New York; 2008

5. Daley, Daniel T.; *Failure Mapping: A New and Powerful Tool for Improving Reliability and Maintenance;* Industrial Press, New York; 2009

6. Daley, Daniel T.; *Reliability Assessment: A Guide to Aligning Expectations, Practices and Performance;* Industrial Press, New York; 2010

7. Daley, Daniel T.; *Understanding the Path to Failure and Benefitting from that Knowledge;* SKF Reliability Systems @ptitude Exchange Article, http://www.aptitude exchange.com, February 2008

8. Ireson, W. Grant & Coombs, Clyde F. & Moss, Richard Y. ; *Handbook of Reliability Engineering and Management— Second Edition;* McGraw-Hill, New York; 1996

9. O'Connor, Patrick D. T. ; *Practical Reliability Engineering— Fourth Edition;* John Wiley & Sons, LTD; West Sussex, England; 2002

Index

www.ingramcontent.com/pod-product-compliance
Lightning Source LLC
Chambersburg PA
CBHW050457190326
41458CB00005B/1321